WORLD INDUSTRY STUDIES 7

The Global
Construction Industry

WORLD INDUSTRY STUDIES

Edited by Professor Ingo Walter,
Graduate School of Business Administration,
New York University

The Global Construction Industry

Strategies for Entry, Growth and Survival

Edited by

W. Paul Strassmann

Department of Economics
Michigan State University

Jill Wells

Bartlett School of Architecture and Planning
University College, London
and
Department of Construction and Environmental Health
Bristol Polytechnic

London
UNWIN HYMAN
Boston   Sydney   Wellington

Published by the Academic Division of
Unwin Hyman Ltd
15/17 Broadwick Street, London W1V 1FP, UK

Allen & Unwin Inc.,
8 Winchester Place, Winchester, Mass. 01890, USA

Allen & Unwin (Australia) Ltd,
8 Napier Street, North Sydney, NSW 2060, Australia

Allen & Unwin (New Zealand) Ltd
in association with the
Port Nicholson Press Ltd,
60 Cambridge Terrace, Wellington, New Zealand

First published in 1988

British Library Cataloguing in Publication Data

The Global construction industry: strategies
for entry, growth and survival.—(World
industry series; 7).
1. Construction industry
1. Strassmann, W. Paul II. Wells, Jill
III. Series
338.4'7624   HD9715.A2
ISBN 0–04–338144–8

Library of Congress Cataloging in Publication Data

The Global construction industry : strategies for entry, growth,
and survival / edited by W. Paul Strassmann, Jill Wells.
    p. cm. — (World industry series; 7)
    Bibliography: p.
    Includes index.
    ISBN 0–04–338144–8 (alk. paper)
    1. Construction industry. 2. Construction industry—
Government policy. 3. Contractors. 4. Competition,
International.
    I. Strassmann, W. Paul (Wolfgang Paul), 1926– .
    II. Wells, Jill. III. Series.
HD9715. A2G57 1988
338.4'7624–dc19   87–32607

Typeset in 10 on 11 point Times by Grove Graphics, Tring, Herts
and printed in Great Britain by Billing and Sons, London and Worcester

Contents

List of Figures

List of Tables

Foreword

International trade in services has in recent years grown in importance, both as a dimension of global economic activity and as a public policy issue. Virtually all of the major industrial countries have seen their service sectors grow rapidly in recent years, both absolutely and relatively, to the point where over half of gross national product in many economies is accounted for by services. This has been reflected in international trade in services ranging from banking and finance, air and sea transport, telecommunications and insurance to consulting and legal services, entertainment and tourism. Governments of countries with an international competitive advantage in these sectors have increasingly pressed for market openings abroad, and the rules governing international trade in services are now on the agenda of the Uruguay Round of trade negotiations under the auspices of the General Agreement on Trade and Tariffs (GATT).

Within a nation, construction is not classified as a service industry, and buildings rarely move across frontiers. Trade occurs in designing, organizing, procurement, management and sitework skills, as well as in sheer labour and equipment use. These transferred construction services find their modern origin well before the turn of the century and are now playing a critical role in major projects in both developed and developing countries. At the same time, access to construction contracts by foreign enterprises raises a number of difficult political and economic issues — as does the presence of foreign construction teams working in host countries.

This volume represents a comprehensive overview of the global construction industry — its structure, the major players and the source of their competitive advantage and the principal policy issues affecting the industry. It will be of substantial interest to practitioners as well as to students of international and industry economics.

INGO WALTER
New York University

Preface

Despite its currently high profile, the international construction industry has never been thoroughly analysed. This apparent neglect may be due to the vast number of different types of project, the multitude of clients, the highly diversified market, and the various nationalities of the contracting firms. To compile information from such a variety of sources and to draw meaningful conclusions from it, would clearly be too big a task for one or two researchers. Hence a decision was taken to enlist the support of others in the compilation of this book. We are most fortunate in having found eleven experts to draw on their extensive knowledge of the overseas construction activities of firms from a dozen countries. After our analysis of American companies and policies, they report on Brazil, Denmark, Finland, France, the Federal Republic of Germany, Italy, Japan, the Republic of Korea, Sweden and Tunisia. Without the wealth of information provided by them in the chapters on specific countries, as well as by correspondence and extensive discussions on numerous points, we could not have attempted our presentations in the introductory and final chapters.

We are also indebted to a large number of others who have helped us with information, suggestions, or critical reading of portions of the manuscript. Particularly helpful on the rise of Japanese firms, apart from our contributors, Yasuyuki Hippo and Saburo Tamura, were Toichi Takenaka of the Takenaka Komuten Company and Toshio Mukai of Kajima International. In the United Kingdom, Walter Allan, David Burgess, Stephen Drewer and Howard Seymour took a detailed interest in the progress of our work and shared their experience, insights, and even work-in-progress with generosity. In the United States, Ingo Walter helped launch the project; and Brisbane Brown, Charles T. Pinyan, and James H. Webber advised on the proper use of that vital ingredient, *Engineering News-Record* (*ENR*) data. For helpful ideas and critiques of major parts of the manuscript, thanks are due to Elias Dinopoulos, Kenan P. Jarboe, Peter Mieszkowski, Lawrence H. Officer, Robert Solo, Diana L. Strassmann, Donald A. Strombom, Josmar Verillo, and David C. Wiggert. On behalf of the authors of all the chapters on specific countries we also wish to express gratitude to the officials of construction firms, research institutes, trade associations, financial institutions, and government agencies. Their cooperation was indispensable. Finally, thanks are due to Maxine Schafer of Williamston, Michigan, for typing the manuscript with great skill and patience.

In support of our nationality-centred approach, it should be noted that overseas contracting firms of particular countries seem to exhibit similar characteristics to one another; and their activities seem to follow observable patterns and trends. Hence it appeared sensible to treat overseas contractors from each country as a group. On the other hand, we are quick to acknowledge a number of weaknesses in the adopted approach. First, although we have tried to cover all of the major construction exporting countries, a number of countries have inevitably been left out. Most significant of these are the UK, Canada, Greece, India, Israel, the Netherlands, the Philippines, Turkey, and Yugoslavia — as well as all of the construction exporting countries of the Eastern bloc.

Secondly, the book deals mainly with contractors, who are responsible for only a part of the construction process: the very important activities of planning and design — functions usually carried out by professional consultants — have not been properly dealt with but only touched upon in connection with design-construct and project management firms. Equally serious is that the viewpoint of developing country clients is represented mainly by Korea, Brazil, and especially Tunisia. There is probably not enough diversity for the probe of several important issues that we attempt nevertheless. In view of these obvious shortcomings, we hope that the book will be read in the spirit in which it was written: as a first attempt to provide some observation and analysis of the patterns and process of international construction.

AUTHOR ?

4420 1 -21
4210
6340

1

Introduction

Strictly speaking there is no such thing as a 'Global Construction Industry' − at least not in the sense that there is a global steel industry or a global textile industry − for 'construction' is not a commodity that can be traded across international frontiers. Each product of the construction industry is in fact fixed to the particular location where it was built. Hence production takes place at a very large number of sites, dispersed throughout the globe, following roughly the pattern of human settlements and economic and social activity.

If the end product is fixed, however, there is evidence that many of the factors of production used in the construction process are exceedingly mobile. A high degree of geographical mobility is indeed required in construction, as the point of production is continually changing. Mobility of resources is also encouraged by the fluctuations that occur in the level of construction activity within any one region or state; surplus resources must be redeployed to new areas when demand slumps and additional resources drawn in when construction activity revives. National boundaries do not seem to present any insuperable barrier to this movement of resources. Thus, there has always been a certain amount of international trade in construction services.

The nature and direction of international trade in construction depends upon the global distribution of construction activity as well as the distribution of wealth. Structures are expensive and can be financed most easily by countries with large savings or the capacity to borrow. Hence the distribution of construction activity is highly uneven, with the richest countries spending upwards of US $1,600 per capita on construction in a year, compared with an average of only US$ 19 per capita in the poorest. Not surprisingly, Japan and the United States constitute the largest national construction markets.

As a result of a high level of construction activity over many decades, all of the rich and highly developed industrialized countries have well-developed construction industries. Although much trade does take place among them, as a group they are self-sufficient in construction materials, equipment and skilled manpower. What they have lacked, especially in periods of boom, is unskilled labour; many of them have become net importers of labour for the construction sector, drawn mainly from less-developed countries. Although the United States and Japan do not import construction labour, others do: Germany imports Turkish labour; France obtains unskilled labour from Spain and Portugal, as well as from countries of the Maghreb; while the United Kingdom has traditionally been supplied with construction labour from Ireland.

Many poor countries, on the other hand, with traditionally low levels of construction activity and little industrial and technological development, have not been in a position to exploit and develop their indigenous resources to build up their construction capacities. Hence, when they do acquire finance for modern construction, they find that they lack the skills and know-how, not to mention equipment and materials, for the efficient execution of construction projects. These have to be imported. Frequently such construction resources are imported in the form of a 'package' provided by a foreign or 'international' contracting firm. International contracting is what this book is all about.

The Rise of International Contracting

International contracting – or firms from one country building under contract in another – is not a new phenomenon. A vanguard of European firms started building abroad with new technology and with the expansion of empire in the nineteenth century. Thomas Brassey, employing English and Irish workers, built the Paris–Rouen railway in 1841, followed by others in Belgium, Denmark, Australia, Canada, Argentina and India by 1860. The French built the Suez Canal (1859–69); and the Germans completed 1,200 miles of an Ankara-Baghdad railway (1899–1914). By that time, Americans were building hydroelectric power stations and petroleum refineries around the world, although the outstanding US overseas construction venture of those years was the Panama Canal (1904–14). The first Japanese project was the railway from Seoul to Inchon in Korea (1897–1900). Energy and transportation projects, many of them with military objectives, were to remain dominant on the international construction scene for decades.

In the 1950s and 1960s, however, international contracting took on new dimensions. Many newly independent states in Africa and Asia launched ambitious development programmes; and the ready availability of loans led to the initiation of large-scale infrastructure, mining, and industrial development projects. The demand for new construction work in third world countries grew at a steady pace. Local construction industries were clearly unable to cope either in qualitative or quantitative terms. Hence, a number of European and North American firms began to expand their overseas activities to meet the increase in demand. For few firms did overseas projects predominate at this stage, however; the majority seem to have used overseas activities merely as a bolster for home country earnings. Some 5 to 10 per cent of turnover overseas was typical.

In the Middle East, the pace of construction activity was also accelerating in the 1960s, following the greater exploitation of oil deposits. Early moves towards modernity included the construction of water desalination plants, electric power stations and improved housing. These were followed by a steady increase in the construction of roads, ports, airports, telecommunications, hospitals and schools. Shortly afterwards industrial development began, and by the early 1970s, petro-chemical complexes and small import substituting industries were common. Then in 1973, the first round of oil price increases brought with it, virtually overnight, a four-fold increase in the revenues of the oil-rich states. Much of this new-found wealth was to be invested in greatly expanded domestic development programmes with heavy emphasis on the creation of new industries. This investment called for an expansion of construction activity throughout the region on a scale and at a pace that was unprecedented. By the end of the 1970s, the Middle East region had become the third largest construction market in the world (after the US and Japan), with an annual output estimated by one author at $150 billion (1982 dollars) (Zahlan, 1983, p. 32). But with little locally available manpower, materials or equipment, and with few local contracting firms, this enormous market was thrown wide open to international competition.

A number of contractors, as well as design consultants, from Western Europe and North America were already working in the Middle East in 1973. Other firms from these countries, who had not ventured overseas before, began to do so. Both groups rapidly expanded their activities (not only in the Middle East, but in Africa, Asia and Latin America, where the demand for new construction work grew as a result of the recycling of petro-dollars). Expansion overseas was paralleled by falling workloads at home, resulting from the marked slow-down in construction activity in the OECD countries after 1973. Hence, for a number of firms, overseas activities came to dominate, instead of merely supplementing home earnings.

The potential gains from overseas construction work (particularly the opportunity it afforded to accumulate capital and to generate foreign exchange), attracted a number of newcomers into the field −

firms from countries which had not been traditional construction exporters, at least not on a significant scale. These included Japan, as well as a number of newly industrializing countries which had already developed a substantial construction capacity of their own. Most dramatic was the rapid overseas expansion of contractors from the Republic of Korea and their capture of a large part of the construction market of the Middle East, a movement strongly fostered by the Korean government.

World GNP (excluding East European non-market economies) amounted to US $12,171 billion (1982 prices) in 1981 (World Bank, 1983, pp. 148–9). About 2 per cent of this output was produced by 250 leading construction companies ($274 billion). Slightly less than half of that work was organized in the home countries of these firms, and slightly more than half ($143 billion) was built in other countries (*Engineering News-Record* (*ENR*), 15 July 1982). If we may call that 'international construction', its share of world GNP was 1.2 per cent and of world construction about a tenth. Regional variations were great and statistical distortions abound, as we shall see, but these figures suggest the overall magnitude.

A Sudden Decline with Side-Effects

In the past few years the international construction market has changed yet again. In 1983 there was a dramatic fall in the value of new contracts awarded to foreign firms in the Middle East, and the downward trend has continued in each subsequent year. This decline is a reflection of a number of factors. First, the sharp downturn in world oil prices, together with a fall in the demand for oil due to the international recession, led to a decline in oil revenues, which in turn had a negative effect upon development expenditure; Saudi Arabia's development budget for 1985–90 was cut by 20 per cent. Secondly, signs of a saturation of construction demand recently appeared in some countries of the region; most of the infrastructure and basic industrial installations are now complete. And finally, in a number of countries indigenous firms have now gained sufficient technical knowledge to compete with international contractors in all but the most complex and specialized fields. The Middle East construction market is no longer what it was; as one contractor complained 'the mega-million pound jobs are not around any more' (*Financial Times*, 12 April 1985).

Foreign contract awards in other developing regions have also declined in number as a result of economic recession and the increasing burden of debt in many third world countries (particularly in Latin

America). The fall in foreign contract awards in constant dollars between 1981 and 1986 was 68 per cent in the Middle East and 50 per cent in other regions with a predominance of developing countries. The international construction boom of the 1970s has clearly come to an end. Almost everywhere the picture is one of shrinking workloads. For example, the value of overseas contracts awarded to firms of France and the Federal Republic of Germany dropped by more than half from 1981 to 1986 while those of Britain and Italy fell by a third (*ENR*, 15 July 1982 and 16 July 1987). Although the mature contractors from the United States and Europe complain about competition from the newcomers, countries relatively new to international contracting have been hit harder. The Koreans have been particularly badly affected by the cutback in the Middle East market and so far appear to have had little success in their attempt to diversify into other areas besides Asia. As a result of drastically declining workloads, a number of Korean firms got into financial difficulties in 1984 and had to be bailed out by their government (see T. Povey: 'South Korea; Government helps ease collapse' in the *Financial Times*, 12 April 1985). Others were compelled to sell out or merge.

The recent spectacular decline in the volume of new construction work in developing regions has had a number of significant side-effects. First, it has led to the development of a buyers' market, with too many firms chasing too few contracts. As a result of this situation, contracting firms anxious to obtain work are frequently obliged to arrange soft credit, involving the provision of long-run concessionary loans at low rates of interest. Thus, the provision of finance has become a major issue of tender. Increasingly contractors find that it is the financing of the project that is demanded of them when they make their initial bids; the best financial package offered seems to have been the criterion for success in bidding in about 7.7 per cent of cases in 1986 (*ENR*, 16 July 1987).

The vast sums of money involved in the financing of construction projects have led many governments to back their contractors with export credit guarantees, aid packages and soft loans. The extent of home government support to contractors obviously varies. Japan and France have been singled out as having not only a high level of government support, but also a highly co-ordinated approach between government and contractors, including extensive help from the domestic banking sector. The use of 'mixed credits' in the support of overseas contracts is a practice which is highly controversial; it has been much criticized, in particular by the United States. Nevertheless, construction industry lobbies are strong enough and the presumed benefits are sufficient to ensure that governments will continue with such practices.

Finally, individual contractors appear to have responded to the dramatic decline in their traditional overseas markets in developing countries in a number of ways. Some have diversified their activities into areas other than contracting; others have attempted to find new markets in developed countries. In the last few years there has been a noticeable penetration of the US market by the Japanese, as well as by firms from a number of European countries, notably Germany, Britain, France and Scandinavia. Similarly, the Italians have increased their activities in other European countries. By 1985, almost one third of the total of overseas contracts won by the top 250 international contractors was in developed regions – North America, Europe and Australia (*ENR*, 17 July 1986). The pattern of international trade in construction services is clearly changing.

Thus, while the Middle East construction boom of the 1970s may well have been a temporary 'one-off' phenomenon, the dramatic expansion and subsequent decline there set off a number of developments in its wake. The situation in the international construction market in 1988 is not back to where it was before 1973. First, the market is no longer the exclusive preserve of firms from the developed world; contractors from Third World countries have entered the arena and shown that they are a force to be reckoned with. Second, there have emerged for the first time a group of genuinely multinational construction firms; firms for whom overseas activities are no longer peripheral, but central to their operations. These firms are constantly seeking out new markets and new investment opportunities, in developed as well as developing countries. Hence it is possible today, for the first time, to conceive of construction as an international industry. As such it is a subject which is currently receiving a great deal of attention from governments, international development agencies, academics, and the public at large.

Measurement Problems and Recent Trends

The value of overseas contracts awarded to the top 250 international contractors has been published each year since 1980 (for the top 200 firms, since 1978) by *ENR*. The information is supplied to *ENR* by the firms themselves, in response to a questionnaire. Some non-responding firms are ranked according to overseas contract awards reported in *ENR*'s weekly newsletter, *International Construction Week* (*ICW*).

The total value of new overseas contracts awarded to the top 250 firms, as reported by *ENR*, for each year from 1980 to 1985 is shown in Figure 1.1. In order to counteract the effect of rising prices, the

Figure 1.1 *Total Value of Overseas Contract Awards to Top 250 International Contractors* (US $ billions, 1982 dollars)

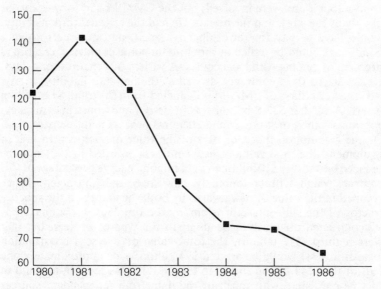

Source: *ENR*, various years.

figures have been converted to constant (1982) dollars. Included in the totals are prime contracts, a firm's share in joint ventures, subcontracts, design-construct contracts, and the erected value of construction management/programme management (CM/PM) contracts. Hence there is clearly a substantial amount of 'double-counting', as much of the work may in fact be subcontracted, all of it in the case of CM/PM. Also, design and construct contracts include the value of installed equipment. Thus the total value of overseas contracts awarded to the top 250 international firms, as reported in *ENR*, is seriously inflated.

An idea of the extent of this inflation can be gained from an analysis of the figures for 1981. In that year the total reported value of overseas contracts was US $143 billion (1982 dollars). Of this, $22 billion was in CM/PM contracts – a sharp increase from the $15.9 billion recorded in 1980. Since the CM/PM firm, in effect, shares work with the prime contractor, if there is one, the gross value of output, though twice reported, should be counted only once.

CM/PM awards are, however, counted by *ENR* only if the firm is exposed to a financial liability similar to that of a general contractor, meaning more than US $500,000. Such exposure applied to over 95 per cent of international CM/PM work since very few worked on a simple fee basis. Cost overruns were still the responsibility of prime contractors if any were involved, but the CM/PM had to warrant the quality of the work and the design. He had the residual responsibility, would have to pay for reworking if needed, and post performance bonds that could be called at any time in some countries, creating a problem of dealing with local courts. A variety of penalties for delays, etc., is set forth in contracts, as well as the limits to liability, if any.

If the $22 billion of CM/PM is deducted from the total, $121 billion is left. A further $37.8 billion was in design and construct (turnkey) contracts, much of it power and chemical process plant construction. On the assumption that a quarter of the value of such contracts is in equipment, the total value of new contracts awarded in 1981 may be reduced by another $10 billion to $111 billion. A large percentage of this contract value will undoubtedly have been subcontracted, hence counted in the value of new awards by both the main and the subcontractor (if the subcontractor is among the top 250). How much was subcontracted obviously depends upon the type of project, but if it were a third ($37 billion), the total value of overseas construction awards in 1981 would be around $74 billion − or approximately one half of the $143 billion shown in Figure 1.1. This reduction would in fact be consistent with much of the data from independent sources presented in Chapters 2–10, which also seems to suggest that the value of new contract awards reported in *ENR* is too high by a factor of two.

Despite the inflated totals, the general trend in overseas contract awards in recent years, as shown in Figure 1.1, is quite clear. There was a sharp increase up to 1981, followed by an equally dramatic fall. The inclusion of an expanded volume of CM/PM awards in 1980 and 1981 obviously serves to exaggerate the steepness of the ascent and the height of the peak. But it does not alter the basic pattern. By 1984 the value of overseas contract awards had fallen, in real terms, below the level of 1979. If CM/PM contracts were excluded, it would be even lower.

Another issue is the relation between awarded contracts and output or turnover. In the national accounts, construction output is not counted as part of capital formation until projects are complete (anywhere from two to five years later). Attempts to relate national investment data to earlier contract awards have, to our knowledge, had little success. On the other hand, turnover refers to the receipts of contractors in any given year, and this figure does seem to be closely related to contract awards two years earlier if allowance is made for inflation. Design changes do occur, raising the final price sometimes much above the initial tender price. The American Bechtel Group, for example, reports the cost of escalations like design changes to *ENR* but does not count that as 'New Work Award Revenue' in its annual reports, thus reducing the total by 9 per cent for 1980–85. More important is that awarded contracts are almost invariably executed. Before those awards are made, projects have often been under preparation for a decade. Nothing is signed until subcontractors and finance are all in place. Only catastrophic economic and political changes – which do occasionally happen – lead to cancellation of contracts.

The lag between contract awards and expenditures (hence receipts) is generally two years. Six to eight months normally elapse while a contractor completes plans, and most will say that the real work begins only in the second year. Spending then speeds up quickly, and before the end of the third year, the bulk of the revenue will have been received. Familiar types of project, including petroleum refineries, will have been completed within two years.

The country contributions to this book, especially German data, confirm the two-year lag between awards and output. The approximate two-fold overstatement of international contracting values by *ENR* can be seen clearly in the American, Japanese, and Italian chapters dealing with countries which, by contrast with Korea, are strong in design-construct contracts.

The Regional Distribution of Contracts

The regional distribution of overseas contract awards for each year from 1980 to 1985 is shown in Figure 1.2. Immediately apparent is the predominance of the Middle East. In its peak year of 1982, the region accounted for 42 per cent of total of international contract awards. Contracts in the region declined dramatically after 1982 (accounting for a large part of the fall in the total), and by 1986 it was no longer the largest single regional market for international construction firms.

The second largest regional market in 1981 and again in 1983 (although way behind the Middle East) was Africa. The peak in contract awards in Africa in 1983, however, can be attributed largely to one project: a massive water transmission scheme in Libya, known as the 'Great Man-Made River Project'. Valued at 3.3 billion dollars, this project has tended to obscure the real trend in the African market (particularly sub-Saharan Africa), which is downwards from the peak reached in 1981.

The Asian market, by comparison, appears to be the most buoyant. In 1984 and 1985 it was second only to the Middle East in the value of new contract awards, and by 1986 it was first with 15.1 billion

Figure 1.2 *Regional Distribution of Overseas Contract Awards, by Value* (US $ billions, 1982 dollars)

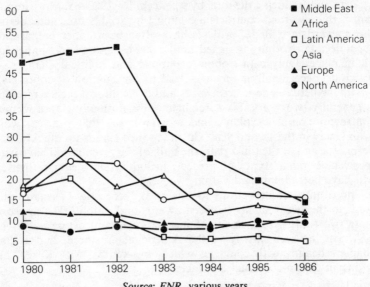

Source: *ENR*, various years.

(1982) dollars. This strength is due in part to the high economic growth rates experienced by several countries in the region and to the availability of multilateral finance for construction projects; and also in part to the opening up of the Chinese market to foreign contractors and investors. It should be noted, however, that apart from Japan two developed countries, Australia and New Zealand, are included in the Asian region. Construction contracts awarded to foreign firms operating in these two countries rose sharply in 1985 and fell as much in 1986. They rose to $4.9 billion − nearly one third of the total value of contracts awarded in the region as a whole − and then fell back to $2.2 billion.

The Latin American market, which is the smallest among the developing regions, accounting for only 7 per cent of the total of overseas contracts in 1986, has also declined steadily since 1981. This fall is a result of the increasing burden of debt and the related economic recession that severely limit the domestic funds available for development projects and led to a marked reluctance on the part of private banks, foreign governments and international development agencies to lend. Neither Latin America nor Africa, where much basic construction work remains to be done, can afford to pay.

The industrial regions of Europe and North America each

constitute about a seventh of the international construction market. However, the value of overseas contract awards in these regions has remained relatively stable, compared with a sharp decline in developing regions. In fact, in the United States there has been a noticeable increase in the past few years in the value of contracts awarded to foreign firms (from $4.8 billion in 1982 to $6.5 billion in 1985). Hence the percentage share of total overseas construction accounted for by Europe and North America has increased, from 13 per cent in 1981 to 30.2 per cent in 1986. If overseas construction awards in Australia and New Zealand are also included, then in 1986 a third of the total value of overseas contract awards were for work in advanced countries. Practically no outsiders built in Japan.

To date there seems to have been little penetration of these developed country construction markets by firms from Third World countries; the overwhelming majority of contracts have been awarded to firms from other developed countries. In 1985, 55 per cent of overseas contract awards in the European market went to other European firms and 42 per cent to firms from North America; the Japanese won 2.2 per cent, and 'others' a mere 0.4 per cent. Similarly, in the North American market, 23 per cent of overseas contracts were awarded to North American firms (Canadian firms building in the US and US firms in Canada); 54 per cent were awarded to European firms (German, French and British, in that order); and 20 per cent to the Japanese. The attempt by the Koreans to break into North American markets has so far met with little success.

The changing regional distribution of overseas construction contracts, particularly the trend towards increased penetration of the construction markets of industrial countries (by firms from other industrial countries), raises a number of interesting questions. First, what kind of service are foreign contractors able to sell in countries which already have highly developed construction industries of their own? The risks involved in undertaking a construction contract are great, as everyone familiar with the industry knows (see the concluding chapter). The risks of building abroad (where contacts are weaker and the institutional framework less well understood) are very much greater. Consequently, would not the prices quoted by foreign contractors be in excess of those of the local industry? It is of course possible that some are selling very special 'high-tech' products or processes that are not available to 'host country' firms; but in view of the rapid dissemination of technological innovations in the world today, this does not appear to be very likely. A possibility might be that there are special circumstances − in the form of political or economic ties − that afford contractors of a certain country an *entrée* into the construction market of another. On the other hand, it may be that contractors working overseas in other developed countries are not simply selling construction services, in response to a pre-conceived demand, but are financing their own projects through involvement in real estate or speculative development. The exact nature of and rationale for such involvement, as well as the precise way in which market entry is effected, are questions which we shall explore in the following chapters.

Changing Shares in a Declining Market: the Advanced Countries

Figure 1.3 shows the distribution of overseas contract awards for seven years (for which comparable data were available), according to the area of origin of the contractor. It is immediately apparent that, despite the entry of firms from developing countries during the boom years of the 1970s and 1980s, the international construction market is still dominated by the industrialized countries. Between 1979 and 1985, US firms captured on average 36 per cent of the total value of overseas contract awards, as reported in *ENR*. A further 39 per cent (average) was awarded to firms from West European countries. In 1986 the American share fell to 31 per cent and the European rose to 46 per cent. And between 4 per cent and 14 per cent went to the Japanese. Firms from these three areas taken together were responsible for 89 per cent of contracts awarded in 1979. Although their dominant position was eroded slightly in the early years of the 1980s, when almost 20 per cent of the market was taken by firms from other areas (including Korea, Turkey and Brazil), in 1985–6 their combined market share had again risen to 89 per cent.

It appears from the data presented in Figure 1.4 that, despite year to year fluctuations, the US share in the world market has remained remarkably stable: in 1985, at 34.6 per cent, it was almost the same as in 1979 and 1981. However, the predominance of the US (and to a lesser extent some West European countries) in the international construction market as reported by *ENR* is almost certainly exaggerated, for the same reasons that the total value of overseas contract awards is inflated. First, US firms undertake a high percentage of the total of CM/PM awards (90 per cent in 1980 and 86 per cent in 1981, the only years for which worldwide data are available). We have already seen that the value of such contracts rose sharply in 1980, inflating US overseas contract awards in that year by $14.3 billion, and by $18.9 billion in 1981. If this kind of contract is excluded from the data altogether, then the share of US firms in the total world market would have been lower in 1980 and 1981. By 1985, however, the value of CM/PM contract awards to US firms had fallen to $6.1 billion (at

Figure 1.3 *Distribution of Overseas Contract Awards by Origin of Contractor* (US $ billions, 1982 dollars)

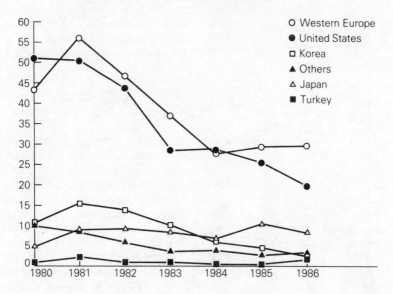

Source: *ENR*, various years.
Note: 'Others' = Brazil + others, mainly Third-World countries.

1982 dollars), and their share in the US work from 36.7 per cent in 1981 to 24.2 per cent (*ENR*, 17 April 1986, p. 92).

A further factor serving to heighten the American share of the market is the high proportion of power and process plant construction contracts that are awarded to US firms, a part of the value, perhaps a quarter, of which, lies not in construction but in installed equipment. In 1981, the value of such contracts awarded to the top 25 firms operating in this line of business amounted to $29 billion, of which US firms won $21.6 billion ($22.9 billion in 1982 dollars), or 74 per cent of the total. This amount represented 45 per cent of the total of all overseas contracts awarded to US firms in that year. Almost all of the rest were awarded to European firms (*ENR*, 15 July 1982). Power and process plants were not reported separately in later years, but the more general category of design-build fell from 51 per cent of US contracts in 1981 to below 30 per cent in 1985. By 1986 the 21.4 per cent US share in design-construct was less than its 30.6 per cent share of all international construction.

Nevertheless, despite these distortions there is no doubt that the US

Figure 1.4 *Percentage Distribution of Overseas Contract Awards by Origin of Contractor*

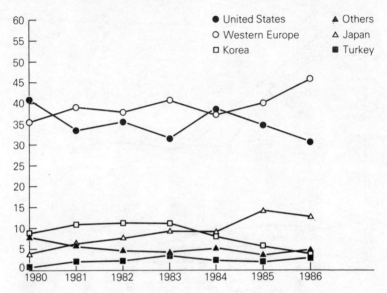

Source: ENR, various years.
Note: 'Others' = Brazil + others, mainly Third-World countries.

is still the largest single exporter of construction services. In 1985 new awards (including CM/PM contracts) amounted to $25.3 billion or 34.6 per cent of the total market. Despite a 22.1 per cent decline in 1986 to $19.7 billion (1982 dollars), the US easily retained a market share twice that of Japan, the second country. The American work was fairly evenly distributed throughout the various regions, with roughly a quarter each in (1) the Middle East, (2) Asia, (3) Africa and Latin America and (4) North America and Europe (see Table 1.1).

The relative shares of other countries in what has been, since 1981, a declining world market are shown in Figures 1.4 and 1.5. The Japanese share in overseas contract awards has in fact increased rapidly – from 4 per cent in 1980 to 14.2 per cent in 1985, and then it fell back to 12.7 per cent. What is more, the real value of these awards rose from $4.9 billion (1982 dollars) in 1980 to $10.4 billion in 1985 – before falling back to $8.2 billion in 1986. Even that partly reflects the 40 per cent rise in the yen–dollar exchange rate. Despite their apparent success overseas, however, Japanese firms among the top 250 international contractors are still the least dependent upon overseas contracts (19.5 per cent in 1985 and 13.2 per cent in 1986 compared

Table 1.1 *Geographical Distribution of Construction Contract Awards, 1982 and 1986*
(All figures are percentages, with 1982 figures in brackets)

Contractor Nationality	Location of Foreign Work					
	Europe and North America	Middle East	Asia[a]	Africa[b]	Latin America	TOTAL
Europe	31.1 (20.4)	13.6 (33.1)	16.6 (11.6)	29.4 (22.3)	9.2 (12.5)	100
US	36.3 (22.9)	34.1 (41.2)	18.1 (20.9)	3.5 (6.2)	8.0 (8.7)	100
Japan	28.7 (3.2)	8.5 (26.9)	57.4 (60.2)	4.3 (8.6)	1.1 (1.1)	100
Korea	3.8 (0.7)	46.1 (77.5)	34.6 (17.4)	15.4 (4.3)	* (*)	100
Turkey	— (—)	50.0 (70.3)	— (—)	50.0 (29.6)	— (—)	100
Others	23.5 (3.4)	20.6 (37.2)	38.2 (11.9)	14.7 (38.9)	5.9 (8.5)	100
All	30.2 (16.6)	21.8 (41.6)	23.4 (19.1)	17.7 (14.4)	7.0 (8.4)	100

Source: ENR, 21 July, 1983 and 16 July, 1987.
Note: An asterisk means less than $50 million.
a Includes Australia and New Zealand.
b Includes Libya and other countries in North Africa.

with an international average of 40 per cent) – which was clearly a reflection of the buoyancy of the domestic construction market.

The regional distribution of Japanese construction contracts overseas, in 1986 can be seen from Table 1.1. Although in 1986 the Asian region still accounted for 57 per cent of the total value of contracts awarded (as might perhaps be expected), it is interesting to note that dependence on the Middle Eastern markets has declined quite noticeably since 1982. This decline has been counteracted by a significant increase in the share of new contracts awarded in North America. The apparent ability of the Japanese to diversify into new regions must be responsible for a large part of their success in international construction markets in recent years. The Japanese chapter will show what lies behind this ability and will answer a number of questions. For instance, have they also diversified into new areas of work? How has their penetration of the North American market been effected? And what exactly are they doing there?

After the US and the Japanese, a number of Western European countries occupy prominent positions in the international construction league. In 1980, France and Germany were the undoubted leaders among the Europeans, with approximately 8 per cent each of the world market; Italy and the UK were some way behind, with 5.9 per cent and 4.7 per cent respectively. Since 1980, the relative positions have changed, however. The German share has declined steadily throughout the period; while the French share dropped dramatically in 1984 but recovered somewhat in 1985 and more in 1986, reaching 9.6 per cent. The UK and Italy, on the other hand, have both managed to increase their market shares (see Figures 1.5 and 1.6). The British rose (from 4.7 per cent in 1980 to 6.9 per cent in 1985 and approached the French with 9.5 per cent in 1986. Italy increased its share of the market from 5.9 per cent in 1980 to 10.0 per cent in 1986, putting it ahead of all other European countries and third in line after the Japanese. The factors that lie behind these recent successes in the world market will be explained in the chapter on Italy.

Despite the changing positions of individual countries, European contractors as a whole have maintained their share of overseas contract awards in recent years. In 1986 they obtained 45.5 per cent of the total, compared with an average of 39 per cent over the previous six years. As with the Japanese, a large part of this success must be attributed to geographical diversification. In 1986, 31.1 per cent of European overseas contract awards were in other European countries and North America, compared with only 20.4 per cent in 1982. Contracts in the Middle East, on the other hand, declined from 33 per cent of the total in 1982 to 13.6 per cent in 1986 (see Table 1.1).

Figure 1.5 *Percentage Distribution of Overseas Contract Awards by Origin of Contractor (European Countries)*

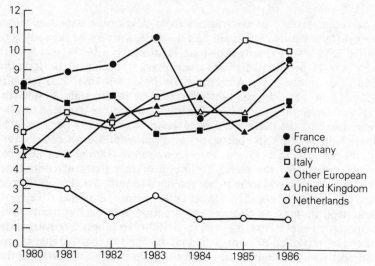

Source: *ENR*, various years.

Figure 1.6 *Distribution of Overseas Contract Awards by Origin of Contractor (European Countries) (US $ billions, 1982 dollars)*

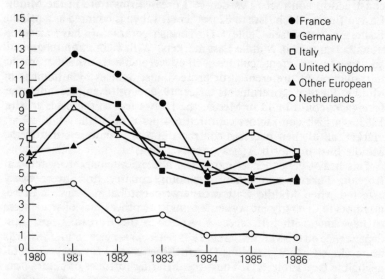

Source: *ENR*, various years.

The Latecomers: Last In, First Out

The recent history of contractors from developing countries tells a very different story. The rapid decline of the Koreans has already been noted, and is very clearly shown on Figure 1.3. In 1983 they captured just over 11 per cent of the overseas market, ahead of the Japanese and second only to the Americans. By 1986 their share had fallen to 3.5 per cent, and they had been overtaken by Japan, Italy, France, the UK and Germany. The fate of Brazilian contractors has been similar; their share of the market fell from 1.5 per cent to 0.3 per cent. The share held by Turkish contractors also fell from 3.6 per cent in 1983 to 1.9 per cent in 1985, but then recovered to 3.0 per cent in 1986.

The reason for the recent decline in market share of contractors from developing countries is not too hard to find. The key to the success of the Koreans (and to a lesser extent firms from other developing countries) in winning overseas construction contracts in the 1970s is commonly believed to have been their ability to supply large quantities of cheap, unskilled or semi-skilled labour, and to manage that labour efficiently with the minimum of impact upon the host country. Hence, they concentrated upon relatively low technology civil-engineering and labour-intensive building projects; and most of their overseas contracts were won in parts of the Middle East which lack a sufficient domestic supply of labour. In 1982, 78 per cent (by value) of overseas construction contracts awarded to Korean firms were in the Middle East; a figure which rises to 82 per cent if Libya is counted as a Middle Eastern country (see Table 1.1). Turkish contractors have an additional advantage in Middle East markets. With a common religion – and to a lesser extent, culture – they were able to benefit from the introduction of preferential or protectionist policies by a number of Middle Eastern governments after 1980. For example, no other foreigners could build in Mecca, the Holy City. In the peak year of 1983, Turkish contractors captured 6.4 per cent of the Middle East market; all of their overseas contracts in that year were either in the Middle East or North Africa (mostly Libya).

This heavy dependence upon the Middle East market meant that Korean, Turkish and other developing country firms were badly affected when Middle East orders were cut back. What is more, attempts to diversify into other regions, and other areas of work seem to have met with little success. Although the Korean 71 per cent dependence on the Middle East in 1985 fell to 46 per cent in 1986, awards in developed country regions (North America and Europe) were still negligible (see Table 1.1). Thus, the differing fortunes of contractors from industrialized and developing countries in the 1980s could be explained quite simply by their ability or inability to diversify their operations into new regions. Less dependent on the Middle East to start with, firms from industrialized countries were able to expand their operations elsewhere, but the others were not.

One reason why developing country contractors are less able to diversify into new areas may be that they are technically less sophisticated and less experienced than their industrialized country counterparts; they do not have the full range of technologies that might be required in international construction or the expertise to compete with developed country firms in certain specialized areas. Another reason may be that they do not have the financial resources required to back up bids in countries where liquidity is a problem or to fund their own construction projects where that is the way into the market.

On the other hand, well-financed international construction projects still come up (according to *ENR* only 6 billion dollars worth of contracts involved some degree of contractor-financing in 1985 and 1986). And many projects do not require highly specialized or complex construction techniques. How many of these are no longer contracted out internationally? Korean (and other) firms have shown that they are extremely competitive where price is the criterion for bidding, and where cheap finance is not a disguised way of cutting the price. In today's construction market, with a large number of firms supposedly chasing a very small number of contracts, one would expect lowest cost producers to be relatively successful. That this is not the case would seem to indicate that price is *not* an important criterion in the evaluation of bids or that the market is structured in such a way that firms from Korea, and other developing countries, are not able to gain access to tender lists. With the information from the country chapters, our conclusion will explore these and other questions.

The Tasks of This Book

We have shown the rise, fall, and re-direction of international construction in recent years. The chapters on various countries that follow should answer how and why the national industries developed and diversified exports as they did and how they coped with the shrinking market. Authors were asked to adapt a given format to the special characteristics of their countries. Each chapter begins with an account of the early history of construction abroad, some major projects, and the role of leading firms. They show how important international construction has become and why contractors are specialized in particular types of projects and customers. These events — entry, growth, and survival – are discussed from the point of view of both business strategy and government policy. Business strategy means preferred types of contracts, bidding policies, association in joint ventures, mobilization of finance, recruitment of labour, links to other industries, and the types of government support that are sought. Government policy includes trade promotion, tax incentives, the provision of insurance and credit, tied aid, and a variety of controls. Some policy measures are related to the development and transfer of technology – hence changing patterns of competitive advantage.

Following the nine chapters on eleven countries is an analysis of their salient differences and similarities. We ask how much more effective government support has been for construction exports in aggressively interventionist countries like France, Japan, and Korea, compared with the moderate official help of Germany, Britain, and the United States. Is financial support really most important, and does that simply come down to a greater willingness to transfer resources abroad?

For the host countries getting technology transferred will in the long run be more important than the resource flows of one year or another. How do builders of different nationalities compare in their capacity and willingness to transfer sitework know-how, design-construct expertise, and organizational skill, especially the sort tied in with sophisticated data bases and computer software?

Both technology transfer and finance have in recent years become associated with joint venturing and other business strategies. Particularly important is the form of contract – design-construct, cost-plus, negotiated lump-sum, open competitive tender, and so forth. Do firms of different countries vary in strategy for becoming short-listed and selected? Are bidding tactics comparable?

Is competition really as fierce as one hears, or should one give more weight to stories about bidding rings, bribes, government pressure, and inside deals? What is more appalling – the risks run by some or the extravagant profits earned by others? Whether firms are competitive or not, can one see an overall secular trend, perhaps going back a half century or more and extending just as far into the future, suggesting how construction exports fit in with a global pattern of development of some countries exploiting technological and economic leads and others trying to catch up?

The chapters about countries, which come next, are arrayed partly in the order of their apparent success in winning contracts abroad in 1985: first the United States, then Japan, Italy, France, the Federal Republic of Germany, and the Republic of Korea. To represent a much larger number of smaller industrial countries building abroad, a chapter on Scandinavia treats Sweden, Denmark, and Finland in detail and briefly comments on Norway. A substantial number of developing countries began building abroad during the booming oil decade (1973–82) and we have chosen to represent these by Korea, Brazil, and Tunisia. The omission of other countries is serious, but so is the alternative, an excessively long book.

2

The United States

W. PAUL STRASSMANN

1 Introduction

The first major and perhaps still the greatest American overseas
construction venture was the Panama Canal. This interoceanic
project was conceived by Alvaro de Saavedra in 1514, attempted
in vain by a French company during 1879–89, and was finally
built by the US Army Corps of Engineers during 1904–14 at
a cost of at least $4 billion in 1982 dollars ($352 million current
dollars). Construction of the canal was finished ahead of schedule,
was unmarred by any reports of graft, and cost less than had
been forecast in money or lives. The labour force of 42,000
mainly West Indian workers were paid a dollar a day (1982
value c. $10) and could do the job because of the virtual elimination
of yellow fever and control of malaria. Nevertheless, 4,500
of the West Indians, plus 1,100 others, died because of the
work.

This chapter will trace American overseas construction from
those beginnings to the mid-1980s. Some further notes about
early ventures and their motivation will be followed by an assessment of the spectacular growth during the 1970s – a near
quadrupling – and the losses of the 1980s. Performance during
the 1980s will be analysed in terms of what has happened to
a set of twenty-three leading firms, and among these, five
giants. American contractors have retained their markets where
complex organizing skills and familiarity with high technology
matter greatly. Business strategies can be seen from the point
of view of preferred contract types, bidding tactics, and relations with
foreign associates. The chapter concludes with a section on the
industry's relation with government, requests for help and protests
against impediments.

The Growth of American Overseas Construction
The American nineteenth century motto that 'trade follows the flag'
probably had it backwards, and it could have been broadened by
another saying that construction contractors follow both. When
planters, miners, and soldiers went abroad, they needed facilities that
local building industries could not produce. To move sugar in Hawaii,
Benjamin Franklin Dillingham built a railway in Oahu in 1889, nine
years before the islands were annexed. Dillingham, Morrison Knudsen
of Boise, Idaho, and other companies had contracts for building naval
and air bases in the South Pacific during the 1930s, and after the
Second World War came more bases and reconstruction. Morrison
Knudsen's first real international venture during the 1940s was a 100
mile railroad in Brazil for shipping iron ore to the new Volta Redonda
steel mill.

Others went to set up complex installations with novel technology,
usually for American overseas investors. Thus the predecessor of
Ebasco built a hydroelectric power system for the American Foreign
Power Company in Chile in 1908; and M. W. Kellogg set up a
petroleum refinery in Argentina in 1926. Bechtel's first joint venture
abroad came in 1940 for construction of a road, telephone lines, an
oil pipeline, and docks in Venezuela. An interesting case involved the
Austin Company of Cleveland, Ohio, that had first gone abroad to
reconstruct Belgian glass plants in 1919. Ten years later under Henry
Ford's direction, they built a Dearborn-like automobile plant at
Nizhni Novgorod, renamed Gorki. The company designed and built
a city for 50,000 and factories that could annually produce 140,000
cars and trucks.

For a decade after the Second World War, some other industrial
nations were too busy rebuilding to give American contractors much
competition in overseas work. Builders from Third-World countries
still lacked capital, skill, and experience. In the late 1960s most international construction was still American. Nevertheless, in the 1960s
the Bechtel Group had only 20 per cent of its contracts abroad (US
Department of Commerce, 1984, p. 21), compared with 65 per cent
during 1980–85. Without including construction management contracts, American construction abroad rose by an annual 14.3 per cent
in real terms from 1970–72 to 1980–82. During that period it nearly
quadrupled from an annual $9.1 billion to $34.5 billion (in 1982
dollars). (US Department of Commerce, 1984, pp. 17, 33). Construction management added $14.4 billion or 46.6 per cent by 1982, making
the annual growth rate 17.5 per cent, but adding some double counting as already explained in Chapter 1. Meanwhile American domestic
construction rose by only 49.5 per cent or at an annual rate of 4.1 per
cent, about that of the rest of the economy. But 1980–82 were the

peak years in international construction, followed by a great slowdown. Before analysing how different types of companies fared during that decline, we will show the extent to which revenues and employment from international contracts actually accrued to the American economy.

Effects on the American Economy

A rather careful and comprehensive survey of output and costs in 1982 and 1983 was conducted by the private accounting and consulting firm, Price Waterhouse, and suggests that gross output may have been half as much as reports of awarded contracts. US gross revenues from overseas contracting were about half as much as *ENR* reports of contracts awarded despite the inclusion of $1.2 billion of billings by designing firms. For 1982, Price Waterhouse reported $21.7 billion of revenues, which compares with an annual average of $49.15 billion (1982 dollars) in contract awards for 1980 and 1981 (which presumably led to the 1982 work and revenues). Of course, *ENR* reports concede that for the same project, construction management contracts, prime contracts, and subcontracts may all be counted again and again. This double counting affects levels more than trends and proportions but, as explained in the introductory chapter, must be kept in mind. Bear in mind also that awarded contracts are often expanded and occasionally cancelled. Incidentally, for international construction, 1982 is as close to an average year as can be found during 1980–85.

How that $21.7 billion was divided among different outlays is shown in Table 2.1. A total of 71.5 per cent was spent abroad; another 3.7 per cent went to American subcontractors; and 12.3 per cent was spent on American materials and components. That leaves 12.6 per cent (one-eighth) for wages, salaries, and fringe benefits of US personnel at home or abroad, for corporate taxes, and for US overhead and profits.

The number of workers involved in exporting construction services under American private auspices (omitting the Army Corps of Engineers, the US Bureau of Reclamation, etc.) was 162,000. Of these 52,500 or about one-third were Americans, but only 22,500 of them worked abroad. Their wages and salaries were about the same at home or abroad if they were permanent employees of a firm. If they were specially contracted for a project, their pay was over a third higher than that of permanent workers. (Table 2.2, lines 2 and 6.)

Foreign direct labour was paid about a third as much as American workers in wages, salaries, and fringe benefits combined. That level may seem high considering the low-income countries where projects were executed, until one recalls that it includes many expatriate

Table 2.1 *Distribution of Costs of Exports of Construction Services, 1982*

		Cost ($ millions)	Share (%)
1	Total revenues (includes designing firms)	21,738	100.0
2	Foreign expenses	15,531	71.5
	a. Materials, subcontractors, taxes and services	(13,994)	(66.4)
	b. Direct and off-site personnel	(1,537)	(7.1)
3	US subcontractors	810	3.7
4	US materials	2,660	12.3
5	US personnel, salaries, wages, and fringe benefits	2,071	9.5
	a. in the US	(1,142)	(5.2)
	b. abroad	(929)	(4.3)
6	US corporate income taxes	97	0.4
7	Overhead and profits after taxes	569	2.6

Source: Calculated from Price Waterhouse, *The Contribution of Architectural, Engineering and Construction Exports to the US Economy* (Washington, DC: International Engineering and Construction Industries Council, April 1985), pp. 16, B1–B4. Materials, personnel, and other expenditures by US subcontractors are included in line 3. If added to the other items, subcontracting would raise these by about 3.9 per cent.

British, Dutch, and other technicians who were high-paid compared with most host country or imported manual workers. Labour costs made up 16.6 per cent of revenues or $3.75 billion. The third who were Americans received 57.4 per cent of this amount.

Gross output of the construction industry in the United States was about $232 billion in 1982. Even in that domestic recession year construction abroad amounted to less than 10 per cent of the domestic volume – $21.7 billion. Since two-thirds of the employment of international construction consisted of non-Americans, the proportion of Americans involved in international, compared with domestic construction, was low, slightly over one per cent out of four million. What is striking is that output and employment for foreign projects was especially important for some firms, notably the largest, as we shall see in the next section.

Table 2.2 Number and Types of Employees, Payroll and Fringe Benefits, 1982

	Number	Annual pay per worker (dollars)	Annual fringe benefits per worker (dollars)	Total cost (millions of dollars)	Percentage distribution
US Personnel employed abroad and in the US					
1 Employees and manual labour	37,257	30,500	8,600	1,456	38.9
2 Contract personnel	6,987	41,800	11,600	373	10.0
3 Support and overhead personnel	8,241	30,100	8,900	322	8.6
4 Total US personnel or cost	52,485	—	—	2,151	57.4
US Personnel employed abroad (per cent of US)					
5 Employees and manual labour	14,939 (40.1)	30,900	7,600	568 (39.0)	15.2
6 Contract personnel	6,218 (89.0)	43,900	12,100	348 (93.3)	9.3
7 Support and overhead personnel	1,357 (16.5)	29,500	5,900	49 (15.2)	1.3
8 US Personnel abroad, total	22,514 (42.9)	—	—	965 (44.9)	25.7
Non-US Personnel					
9 Direct labour	100,789	14,200	—	1,430	38.2
10 Support and overhead personnel	8,965	18,500	—	166	4.4
11 Total	162,239			3,747	100.0

Source: Calculated from Price Waterhouse, *The Contribution of Architectural, Engineering and Construction Exports to the US Economy* (Washington, DC: International Engineering and Construction Industries Council, April 1985); Appendix B. pp. 2–3. About 3.6 per cent of employment by US subcontractors is included.

2   Structure and Performance in the 1980s

Size of Firm and Participation
Among the largest four hundred American construction firms, over a
quarter (106) won contracts abroad in 1985, sometimes no further
away than Canada or Mexico. Forty-three had acquired enough
foreign work in 1985 and 1986 to be included among *Engineering
News-Record*'s 'top 250 international contractors' (17 April, 17 July
1986 and 16 July 1987). Americans made up one-sixth of the firms and
received 34.6 per cent of the contracts awarded, amounting to $25.6
billion (1982 dollars) as the leading contingent. In 1986 the share was
down to 30.6 per cent, and the amount was 19.7 billion dollars.

The American international contractors fall into a few major and
a number of minor overlapping categories. Foremost are the chemical
and petrochemical process plant constructors like M. W. Kellogg,
Foster Wheeler, Lummus Crest, C. F. Braun, Stearns Catalytic,
Badger, and Arkel. Another category are firms in charge of heavy civil
engineering projects, dams, highways, bridges, pipe lines, sewer
systems, mines, harbours, and airports. Outstanding among these are
Morrison Knudsen, Peter Kiewit Sons, Guy F. Atkinson, and Brown
& Root (especially known for off-shore oil drilling platforms). Some
heavy contractors like Ebasco and Stone and Webster have specialized
in electric power plants of all types, but are now branching out. CBI
Industries and Pitt-Des Moines build mainly a variety of storage tanks
and colossal containers. Another major category is made up by the
general builders like Rust (formerly affiliated with Kellogg), Turner,
Perini, Dillingham, and Austin. They have done best with highrise
buildings in Taiwan, Hong Kong, and Singapore, or with high
technology centres. Finally, there is the category of the huge diversified contractors who construct anything through all their divisions,
subsidiaries, and overseas affiliates. Parsons, Bechtel, Dravo, and
Fluor in 1985 operated in 44, 52, 41, and 22 countries respectively.

The forty-three firms that made *ENR*'s 'top 250' list won $588
million in foreign contracts per firm in 1985 and received 99.6 per cent
of the contracts awarded to 106 American firms. The remaining sixty-
three US firms received 0.4 per cent of all contracts by value, only $1.7
million per firm. Out of a million smaller firms, only a tiny handful
ever worked abroad. The experience of US firms included among the
top 250 therefore seems a good approximation of the overseas work
of American contractors.

The concentration of contract awards to a few firms seems dramatic
if one divides the 43 included firms into three groups — the top 5, the
next 15, and 23 others. The top 5 had 71.3 per cent of overseas

contracts awarded to US firms or $3.239 billion per firm; the next 15 had a third as much, 24.3 per cent or $368 million per firm; and the others only 4.0 per cent or $44 million per firm. The concentration ratio for the top 4, a frequently used term in manufacturing, was 62.7 per cent, higher than the average for the twenty major ('2 digit') American manufacturing divisions: 38.6 per cent (Allen, 1976, p. 665).

Five Leading Firms
The top five American foreign contract winners of 1985 − Kellogg, Parsons, Bechtel, Brown & Root, and Lummus Crest − were also the top five international contractors in the world. Since they had to compete for some $73 billion of 1985 overseas business with European, Asian, and other firms, a purely intra-American concentration ratio does not reflect the extent of competition. Of that $73 billion, the share of these five firms was 24.8 per cent, still astonishing, but not overwhelming. For four firms it was 21.8 per cent.

Although these five mostly worked abroad, purely domestic American contracts nevertheless amounted to 36.5 per cent of their business. In the domestic market, worth $96.6 billion for the relevant types of construction (*ENR*, 17 April 1986), they had to compete with other American firms, including the vast majority who had little or no interest in foreign work. The concentration ratio with foreign and domestic work combined is 25.1 per cent for the 5 or 21.0 per cent for the 4 largest (omitting Lummus Crest). In other words, foreign and domestic concentration was about the same. Abroad, more foreign competitors made up for fewer challenges from other Americans. The extent of competition has to be judged in a more disaggregated manner, at times on a project by project basis, but at least according to this indicator, construction remained fairly competitive.

Oddly enough, the concentration ratio had actually risen from 15.2 per cent in 1981 as competition increased with declining business. Altogether the top 5 rose from having 58.2 per cent more business than the next 15 combined, to having 193.4 per cent more business, that is, to a level nearly three times as high. By 1986 their share was three and a third times as high.

How the top 5 firms compared with the next 18 can be seen in Table 2.3. This table gives the average annual value of contracts won during 1980–85 in constant 1982 dollars. A six-year average is better than any single year's figure since even the largest firms have erratic ups and downs as megaprojects and reorganizations are undertaken in some years but not in others. For example, Fluor had captured the number one spot in 1979 only because of its $2.2 billion contract to build a coal gasification plant in South Africa. Later Fluor withdrew from South Africa.

Table 2.3 *Twenty-three American International Construction Firms
Average Annual Value of Contracts Awarded 1980–85
(1982 dollars, billions)*

Firm	Foreign contracts	US contracts	Total contracts	Percentage of foreign work	Change from peak to trough (%)[c]
1. Bechtel Group	6,072	3,264	9,336	65.0	− 67.1
2. Kellogg Rust[a]	5,150	3,331	8,481	60.7	(+ 43.7)
3. Parsons	4,641	3,219	7,860	59.0	− 71.0
4. Fluor	3,023	2,486	5,509	54.9	− 70.0
5. Foster Wheeler	2,416	940	3,356	72.0	− 65.1
6. Lummus Crest	2,388	1,224	3,612	66.1	− 32.1
7. C. F. Braun	1,983	221	2,204	90.0	− 88.0
8. Brown & Root	1,622	4,224	5,846	27.7	− 33.8
9. Morrison Knudsen	1,401	2,457	3,858	36.3	− 89.0
10. Guy F. Atkinson	1,010	1,015	2,025	49.9	− 95.4
11. Raymond International	967	1,590	2,557	37.8	− 84.2
12. Austin	842	754	1,596	52.8	− 72.0
13. Stone & Webster	622	1,221	1,843	33.7	− 68.1
14. Dravo	533	1,281	1,834	30.2	− 62.2
15. Stearns Catalytic[b]	542	2,522	3,064	17.7	− 39.2
16. CBI Industries	452	446	898	50.3	− 82.1
17. Jacobs Engineering	398	1,146	1,544	25.8	− 97.5
18. Dillingham	284	654	938	30.3	− 43.2
19. Ebasco	239	1,549	2,788	8.6	− 89.1
20. Turner	161	1,943	2,104	7.7	− 85.3
21. Peter Kiewit Sons	139	847	986	44.3	− 12.8
22. Perini	76	983	1,059	7.1	− 92.5
23. Arkel	60	75	135	44.3	− 63.1
24. Average	1,480	1,795	3,276	42.3	− 63.9

Source: *Engineering News-Record*, annual surveys of the 'top international contractors'. Values are converted to 1982 dollars with the GNP deflator. These 23 firms are all those who replied to the *ENR* surveys for each of the 6 years. The first 20 firms are also those with the largest contract awards. The last three are included only because they reported each year.

Notes: [a] Kellogg and Rust merged in 1981 and split in 1985 but are included as Kellogg Rust for all years. Their combined volume was $7,983 million in 1984, $5,637 million in 1985; bigger than the $3,924 million of 1981.

[b] Before 1982 Stearns Catalytic were reported separately as Catalytic and Stearns Roger. For this table, the separate companies were added.

[c] The peak was either 1980 or 1981, and the trough 1984 or 1985.

Among leading firms, the share of construction management in their new foreign contracts of 1985 was Parsons, 84.1 per cent; Ebasco, 39.8; Fluor, 36.5; Bechtel, 34.8; Dravo 25.5; Morrison Knudsen, 7.8; and Brown & Root, 2.5. In Bechtel's peak year of 1980, 77.6 per cent of its $9.85 billion in international contracts was design-construct, with the remaining 22.4 per cent construction management. In 1985 new contracts were down to $3.25 billion with only 17.1 per cent design-construct. With a decline from $7.64 billion to $564 million, 92.7 per cent of the design-construct work had been lost in real terms. Meanwhile design-only projects began to approach an annual billion. Nevertheless, Bechtel employment fell from over forty thousand to less than twenty thousand. Among those with new jobs were President Reagan's Secretary of State and Secretary of Defense.

Twenty-three Firms
The domestic and foreign volume of work of 18 other firms can be compared with the 5 leaders in Table 2.3. Brown & Root was first in domestic work but only eighth internationally. Ebasco was fifth domestically and only nineteenth internationally. For 3 firms international work was less than 10 per cent, but for 5 it was over 60 per cent. For one it was 90 per cent: C. F. Braun of Alhambra, California, now a subsidiary of the Kuwait National Petroleum Company. On the average the foreign share of these 23 firms was 42.3 per cent.

The top 20 firms in the table are included because they were the largest American international contractors during 1980–85. Three others are included because they were the only other ones reported among the *ENR* top 250 for each year during 1980–85. Some firms with occasionally larger awards were erratically in and out because of mergers, absorption, failure to report, severe fluctuations, and perhaps other factors. The Badger Company of Cambridge, Massachusetts, a chemical process plant builder and a subsidiary of Raytheon since 1968, had an average international volume of $323 million during 1980–83 or 17.5 per cent of its business. Badger's international work continued, especially a $1 billion refinery at Whangarei, New Zealand, but for some reason Badger stopped reporting to *ENR*.

Decline and Survival
Although 1981 was the peak year for international contract awards with $143.2 billion, 15 per cent more than in 1980 in real terms, the amount awarded to the constant set of 23 US firms was already declining by 5 per cent from $47.6 to $45.4 billion. In relative terms 1980–81 was the worst decline for American international contracting (see Chapter 1, Figure 1.4). The share of the 23 fell from 38.2 to 31.6 per

cent, and the share of all US firms among the top 250 contractors fell from 40.8 to 33.1 per cent. Trends among the constant set of 23 are a better measure of what is happening since values do not drop to zero for any firm that fails to be included among the 250.

In 1982 tight monetary policy pushed the American economy into a severe recession with 10.7 per cent unemployment and with output 9.5 per cent below the level that would have kept inflation and unemployment constant. Domestic contract awards to the 23 firms had been stable around $55.5 billion in 1980 and 1981, but they now fell by half to $27.4 billion. Since domestic construction fell by more than international construction, the *share* of international work in the order books of American contractors actually rose in 1982 from 45 to 56 per cent (Table 2.4, line 8).

The 23 American firms collectively experienced a 49 per cent decline from their 1980 peak. The sharpness of the descent looks greater if peaks and troughs appropriate for each firm are used. Individually some firms had their peaks in either 1980 or 1981 and their troughs in 1984 or 1985. As can be seen back in Table 2.3, the maximum decline per firm, using peak and trough years, was actually 63.9 per cent. Turner, Braun, Morrison Knudsen, Ebasco, Perini, and Atkinson had declines of 85 per cent or more. Not only had they landed especially large contracts early in the decade, but their specialties of heavy engineering and energy-related structures were particularly vulnerable.

The amount of American international contracting stabilized around an annual $25 billion and thus actually became an *increasing* share of the shrinking international total. But as domestic American construction recovered from $27.4 billion to $39.1 billion, the share of international work for the 23 firms actually fell from 56.0 to 38.1 per cent. Nevertheless, American firms seemed now to be in niches where their competitive advantage was greatest and where they would not be easily displaced. That these niches involve technology and organization will be shown in coming sections.

The viability of American overseas contracting was actually better than the figures of Table 2.4 suggest for 1983–5. The totals for international contracting include contracts won by foreign firms in the United States, an amount that rose from 3.6 to 7.1 billion dollars with the recovery of the American economy during 1983–5. American gains in this market do not, of course, qualify as 'international'. Fully one-third of non-American international contractors, 67 firms, won contracts in the United States during 1985, up from 57 in 1984. If these amounts are subtracted from the international totals of $74.5 and $73 billion for 1984 and 1985 to focus on the non-US international market, then the given American share is about a tenth (3 per

Table 2.4 Recent Trends in the Share of US Firms in International Construction Contracting

	1980	1981	1982	1983	1984	1985	Total six years
1 International contract awards, billions of constant 1982 dollars	124.7	143.2	123.1	90.1	74.5	73.0	628.6
2 US firms among the top 250	48	47	46	43	41	43	82
3 Share of US firms, per cent	40.8	33.1	35.4	31.4	38.1	34.6	35.8
4 Share of subset of 23 firms, per cent	38.2	31.6	28.4	28.3	35.4	33.0	32.4
5 International contracts awarded to the 23 firms ($billions)	47.6	45.4	34.9	25.5	26.4	24.1	203.9
6 Domestic contracts awarded to the 23 firms ($billions)	55.5	55.4	27.4	33.0	36.6	39.1	247.0
7 Total contracts awarded to the 23 firms ($billions)	103.1	100.8	62.3	58.5	63.0	63.2	450.9
8 Share of international contracts in the total workload of the 23 firms, per cent	46.2	45.0	56.0	43.6	41.9	38.1	45.2

Source: Annual surveys of Engineering News-Record. Values converted to 1982 dollars with the GNP deflator.
Note: The total in line 1 contains the value of contracts won by foreign firms in the US.
The 23 firms are those listed in Table 2.3. They are all those that reported for each of the 6 years and usually accounted for about 95 per cent of US international construction.
The greater difference between the 23 and all firms in 1982 was due to the singular appearance of C. T. Main of Boston at the top of the list with $5.2 billion in contracts.

cent) bigger than indicated in Table 2.4, lines 3 and 4 for those years. No corresponding expansion of US construction in Europe or Japan took place during those years to offset that effect. Hence, in the rest of the world, the US held its own and was not losing competitiveness.

In 1986, the volume of new contract awards fell by over a fifth to $19.7 billion (1982 dollars). Design-construct work fell by nearly half, or from 32 to 19 per cent of American international construction, and therefore accounts for the decline. Construction management rose by 14 per cent, and other types of construction by 12 per cent. Design-only work by these companies (not included in the total) also rose by 4 per cent. Construction management had now risen from 19 to 37 per cent of American overseas work (*ENR*, 16 April 1987). How these changing proportions reflect the American competitive advantage will be discussed next.

3 The American Competitive Advantage

Relative Costs
For conventional buildings and well-known types of infrastructure, American construction firms do not have a cost advantage. As the president of the international division of a company specialized in building and heavy infrastructure said during an interview with the author: 'We are as efficient as anyone, but not more so. Civil engineering technology is universal. In on-site construction, everyone knows everything. Perhaps the Japanese are the best in certain novel ways of making tunnels, but in general, no one has much of an advantage.' In the Middle East, contractors of all nations employed Thais, Filipinos, Pakistanis, Portuguese, and others, either directly or through labour contractors. Hence no third country builder, except Koreans and Chinese regimenting their own, had an edge in labour, according to this view.

In 1985 the services of the half dozen senior technicians supervising construction of a power plant in Saudi Arabia were billed at $97,500 per month for each ($87,300 in 1982 dollars). This rate was based on the cost of bringing Americans to the site, but in fact all six were from Europe or the Far East and were paid half as much. That is not to say that the contract price was not competitive. On the contrary, the project manager said in an interview that the firm was the lowest of twenty bidders: 'It's not really wise to be in that position, but we bid that low because we were anxious to stay in Saudi Arabia.'

For relatively conventional projects throughout the world, the high salaries of Americans make their employment overseas prohibitively expensive. The firms working on the foreign-aid-supported Cairo

sewer project could afford to employ Americans for about half of the
supervisory tasks. Their pay made up 55 per cent of *all* wages and
salaries, supervisory and site labour. On many other sites and at headquarters of subsidiaries, however, not one American will be found.
Typically an executive from the home office will visit four times a year
to cope with unusual situations, promotions, and the like. Telecommunications do the rest.

Traditional Projects and Training
The principal American competitive advantage is not in using familiar
building methods for well-known types of structures. It lies in
management and design. Knowledge of a practical character is incorporated in structures and equipment that are novel to the receiving
country. Nowadays technology transfer means not just installation
and demonstration, but teaching the designing and management process itself.

For highways, dams, urban infrastructure, housing, office
buildings, and other traditional projects, American firms are not
needed to transfer technology. Firms specialized in such projects
could win contracts in the Middle East during the peak of the oil
boom; but because of their higher costs, they could no longer meet the
competition in the shrinking market of the mid-1980s. In some cases
their volume fell from several hundred million annually to 10 or 20
million dollars of foreign-aid supported work or in upgrading the
security of US embassies. However, many major American overseas
builders had never been interested in these or other foreign civil
engineering projects.

Exceptions to the decline of traditional work might occur in the
cases of highly complex 'megaprojects', such as Atkinson's Guri Dam
in Venezuela, Brown and Root's two Great Man-made Rivers in Libya
(with Dong-Ah), or the trans-Panama pipeline, the King Khalid
Military City of Saudi Arabia, or the Colombian Cerrejon Coal Project, all built by Morrison Knudsen (M–K). The two billion dollar
Cerrejon project has been the company's largest. This turnkey contract was awarded in 1981 and was completed five years later by M–K.
using some 200 subcontractors, 21,000 suppliers, and a labour force
that reached 11,000. Of these 381 were expatriates. An open-pit mine
is to produce 15 million metric tons of coal annually. Construction
included a 90-mile railway to the coast, a new port with high-volume
transloading facilities, and related infrastructure. An official of M–K,
Vern L. Nelson, told us in early 1986 that, the 'market supports the
conclusion that international clients favour US participation in
engineering, procurement and construction management services, not
the traditional "full responsibility" service as was provided in the

past'. M–K's foreign contract awards were down to 62 million (1982) dollars, a mere 2.8 per cent of its new work, and M–K closed most of its international offices.

Any aspect of a project that will be repeated in others can be taught, and such teaching is now expected by both overseas associates and American firms operating internationally. As Russell J. Christesen, President of Ebasco, wrote to us in 1986:

Technology transfer is rapidly becoming a contractual requirement to enable countries to become either self-sufficient or competitive exporters. As a result, we find ourselves participating in projects of a different nature than our initial thrusts were focusing on. Projects such as desalination of water, rapid transit facilities, hazardous waste storage and disposal, as well as crest of the wave technology in conventional projects.

The director of international operations of another company explained how they were training a young US-educated Libyan to run a billion dollar project:

In three years he will learn it all. We are better at technology transfer than the Europeans. We show them by means of a sort of buddy system. We train, we develop manuals, we explain computer codes, things like that. It's not just the product being built, but showing the process of designing it. We have a nuclear power project in Korea, and they want to compete abroad building that sort of thing themselves. But even with all the technology transfer, they're still a long way from that. Meanwhile, we're developing other designs. Technology transfer is the best thing we have going for us.

This have-technology, will-transfer approach is advertised in the firms' brochures. For example, one of Badger Engineers says the company 'has long been a leader in the design of first-of-a-kind plants and has repeatedly demonstrated the ability to transfer sophisticated technology to its clients throughout the world'.

Chester Lucas, a veteran international engineer and corporate executive, has observed the tendency of both sophisticated clients and international lending agencies to insist on a heavy training component in large design and construction management contracts. But like many of his colleagues, he regards such teaching as a calling, a duty that one should discharge even without contractual obligation. 'Engineers who can train technicians in foreign countries are the unsung heroes of the consulting profession . . . In the end the philosophy of passing on the torch and working oneself out of the job will be instilled in all of the consultant's expatriate staff' (Lucas 1986, pp. 149, 217).

Technological Frontiers and Niches

The Bechtel Group was not only the largest firm during 1980–85, with an average annual international volume of contract awards of $6.07 billion (1982 dollars), but it was also the most research-oriented. Bechtel is owned by its senior managers who must sell stock back when they leave. The company began as a railway builder in 1898; went on to bridges, pipelines, and dams; and around 1940 began design-construct work. By the mid-1980s it had a research and development unit that was unique among construction firms. It had about 300 employees who were encouraged to work on ideas without immediate commercial application. In 1986 they were concerned with lasers, advanced energy systems, superconducting magnets, novel materials, and biotechnology. On the verge of commercial introduction were novel X-ray cargo screening technology, flue gas desulphurization devices, large fuel-cell power plants, hazardous waste-management technologies, fibre-optic cable systems and plans for a factory making integrated electronic components.

Workers throughout Bechtel were encouraged to advance innovative ideas, and winners of annual competitions could hope for research grants of 50 to 100 thousand dollars. A Bechtel executive has reported to us that the company holds about 150 patents and that its employees publish almost 500 technical papers annually, 'a commitment to research that is without parallel in the industry'. Results of past research were being applied in eighteen countries in 1986, the foreign share, of work outside of Canada being down to 30 per cent. Projects included nuclear power plants in Korea, an across-the-Andes pipeline in Colombia, an aluminum smelter in Australia, floating oil storage units in China, an ammonia-energy complex in Sweden, a polymer bottle factory in Britain, as well as hotels, resorts, motorways, airports, mines, and other projects throughout the world.

The second American overseas contractor during 1980–85 was M. W. Kellogg of Houston, who actually led with $5.55 billion (1982 dollars) in new contracts in 1985 and $4.44 billion in 1986, especially for fertilizer plants in Iraq and Australia. Kellogg's work was all design-construct or design-manage, nothing in traditional building or civil engineering. The foreign share of Kellogg's work was 91.8 per cent in 1985 and 73.2 per cent in 1986. Ammonia and liquid natural gas plant patents had put the company ahead in recent years. The company keeps the technological frontier moving at a research centre that does many things from mathematical modelling to tests of indoor pilot plants up to 60 feet in height, but in the 1980s it carried out no research in building techniques *per se*. According to a company official, 'Our proprietary technology in ammonia, catalytic cracking, and ethylene has been critical to our success internationally.'

Another builder of chemical and petrochemical process plants said,

> The Japanese have not yet caught up in this field. This type of process engineering is hardest for others to learn. Each project is different. It's not repetitive. Of course, the supporting systems to our plants are mere textbook engineering, but still we often have had to educate the foreigners in all of that, foundations, steel structure, the layout of piping, and the electrical distribution systems.

In industrial countries this company typically builds 10–15 per cent of the structure (only the key installations); but in developing countries, as much as half was the usual share. That amount could well be worth several hundred million dollars.

These process-engineering, design-construct firms have not been particularly interested in building infrastructure, office buildings, and support facilities related to a project. In good times such work is readily subcontracted, and in hard times these firms have been ready to be the subcontractors themselves, filling the technological niche for a Korean or Japanese general contractor.

The word 'specialty' or 'niche' is often used by executives of technology-based overseas builders. Two of these are Pitt-Des Moines (PDM) and CBI Industries (Chicago Bridge & Iron Company). Often as subcontractors, these firms build specialized storage tanks out of steel, aluminum, or alloy plates. PDM began in 1893 with water towers, but its specialty is now in enormous containers for extremely high and low temperatures and pressures. Skill at designing and fabricating floating yet sealed roofs is critical. Alumina refineries, liquid gas tanks, space simulation chambers, high energy laser facilities, wind tunnels, domes for athletic stadiums are what PDM builds.

One could go on with other firms in other specialties. The Austin Company of Cleveland has built all sorts of factories and commercial facilities, including the world's largest volume structure, the Boeing 747 plant at Everett, Washington. Beginning with lightbulb factories for General Electric in 1908, Austin developed the fast-track concept of integrating design, engineering, and building and applied it to plants for making automobiles, aircraft, newspapers, chemicals, pharmaceuticals, and many other products. It tends to follow the multinationals overseas but will build satellite tracking stations and data processing centres for clients throughout the world, though primarily in other industrial countries. 'We sell American expertise', Russell

Silliman, Vice-President for International Operations, told us. 'High technology centres are our bread and butter.'

How long the American technological advantage will last cannot be predicted. It may be primarily due to the diversity of tasks that modern structures are asked to perform. Each task makes exacting demands on the structure that are not obvious from experience elsewhere. If the diversity of tasks proliferates faster in the United States, so will the demand for novel structures, experience in building them, and a competitive edge over less experienced builders in other countries.

With its large population and high incomes, the United States indeed remains the world's premier market, where the division of labour is least limited, where economies of scale can be realized smoothly, and where both sensible novelties and silly fads can be brought into production earlier than elsewhere. Preoccupation with military security leads to the need for building other novelties. Civilian spinoffs take place now and then.

Organizing Skills
American firms win contracts abroad, not only because of their experience with advanced technology, but also because of organizing skill. Indeed, programme and construction management (PM/CM) contracts are sales of that skill. Some firms like Bechtel were even willing to undertake 'procurement-only' contracts.

The PM/CM firm can be involved during the feasibility and design phase (hence 'project management') and advise on constructability, selection of materials, availability of labour, schedules, and budgets. The PM/CM firm can prequalify and select designers and constructors for the client, check the contractors' equipment, inspect work, disburse funds, and supervise training. Ability to enforce an integrated engineering, procurement, and construction schedule under standard cost control procedures is the essence of CM. Also involved are skill at manpower forecasting, labour conflict resolution, analysis of materials delivery systems, and anticipation of environmental or political effects. Much of this planning has been enhanced by highly sophisticated computer-based data analysis, a field in which the United States has a substantial lead.

Perhaps the world's largest PM/CM firm is the US Army Corps of Engineers, whose 40,000 employees manage not only the building of military installations but a wide variety of civilian programmes in the US and abroad. The Corps' Middle East Division has managed over $20 billion worth of construction projects for the Saudi Arabian and other governments, including television systems, airports, harbours, warehouses, schools, hospitals, housing, shopping centres, recreational facilities, and urban infrastructure. The Corps' undated brochure, *The Middle East Division*, reads like one from any private firm: 'Satisfying customer requirements is this organization's driving force . . . When a Client calls upon the Middle East Division to manage a project, he makes an investment in quality and cost control' (pp. 3, 17). The Corps operates on a cost-reimbursable basis without profit.

The importance of construction management for ten representative American firms is shown in Table 2.5 with data for international construction's peak year, 1981. Some firms like Austin, Kellogg, and Lummus did no construction management. Except for the Rust International, partner of Kellogg, they did nothing but design-build. Others like Burns and Roe Enterprises of Oradell, New Jersey, only do construction management.

As early as 1981, most of the large firms already did both construction management and design-construct but little construct-only, meaning carrying out designs of others after competitive bidding (Table 2.5, column 4). Except for Parsons and Morrison Knudsen all the firms were overwhelmingly in high-technology installations, power stations and chemical process plants (Table 2.5, column 5). Burns and Roe also mainly managed manufacturing process and power plants, and Austin did the other types of installations that have already been described.

Note that only Brown and Root and Parsons did a substantially larger share of construction management abroad than at home (Table 2.5, columns 1 and 2). The other four firms, the Bechtel Group, Dravo, Morrison Knudsen, and Fluor, both abroad and at home averaged 52 per cent of their contracts in CM. By 1985 the share of CM for these firms remained about the same for two (Parsons, 84.1 per cent; Fluor, 36.6 per cent) and fell for four (Burns and Roe to 38.5, Bechtel to 34.8, Morrison Knudsen to 8.0, and Brown and Root 0.4 per cent). The remaining firms stayed at zero. Dravo, according to a letter from an executive of the firm, no longer had any construction management contracts abroad. One cannot, therefore, say that American overseas building was being pushed into nothing but management.

The list of firms in the table suggests that most carry on each of the three major types of work, but that some do only construction management and others only design-construct. We should add that a few outstanding firms retain the traditional approach doing no designing and no construction management abroad but are 100 per cent 'construct only'. One is Perini of Massachusetts who have built dams, radar stations, air bases, urban infrastructure, and highrise buildings all over the world. Another is Guy F. Atkinson of San

Table 2.5 *Share of Construction Management, Design-Construct, Construct-Only, and Power or Chemical Process Plants in Foreign Construction of Selected American International Contractors, 1981 (%)*

Firm		Construction management as share of US construction (1)	Foreign construction			Share of power and chemical process plants in 3 and 4 (5)	Foreign construction contracts awarded during 1981 (millions of 1982 dollars) (6)
			Construction management (2)	Design-construct (3)	Construct only (4)		
1	Burns & Roe	100.0	100.0	0.0	0.0	—	523
2	Parsons	18.7	83.0	14.5	2.4	(a)	3,875
3	Bechtel Group	48.7	61.6	38.5	0.0	79.7	7,341
4	Dravo	54.9	56.8	32.1	11.1	99.8	1,105
5	Morrison Knudsen	58.9	53.3	42.6	4.1	(a)	3,399
6	Brown & Root	0.0	45.4	48.6	6.0	83.9	2,802
7	Fluor	46.9	38.1	59.5	2.4	92.0	4,454
8	Kellogg Rust	0.0	0.0	86.4	13.6	80.0	3,924
9	Lummus	0.0	0.0	100.0	0.0	100.0	2,856
10	Austin	0.0	0.0	100.0	0.0	0.0	364

Source: Engineering News-Record, 15 July 1982. 1981 values were converted to 1982 dollars with the GNP deflator. Data for other firms were incomplete.

Note: Data for Parsons and Morrison Knudsen not provided but the share was less than 50 per cent.

Francisco, especially known for dams, including the Grand Coulee on the Columbia River, still the world's largest, and the Guri Dam of Venezuela, completed in 1986, a $5 billion project. If design-construct is called for, these firms will subcontract the designing part to the 'most appropriate firm'. Their executives say that this strategy is actually an advantage and that they prosper most with 'hard-money and hands on' specialization as builders.

In CM, computers help with processing and storing of information and can give early signals for extra attention to some part of the work due to delivery and schedule problems. But 'Critical Path' and PERT (Program Education and Review Technique) systems for scheduling activities and controlling costs go back to the 1950s. According to old-timers having a computer is much like having an expensive tower crane: it forces one to be well-organized in many ways that were just as possible before acquiring the expensive tool. Executives of firms that have used modern management methods for decades are likely to say things like, 'Computers save labour but cause no revolution', 'Computers don't really add anything new', and 'Any change that comes from computers is all hype; though we even have our own software company, I tell you construction will always be a sweat industry, never doing the same thing twice.' But a younger vice-president thought 'computers are changing our whole corporate culture and may make us cost-competitive again'.

The organizational advantage of old American firms lies partly in their long established network of international contacts. Said the Senior Vice-President for International Operations of one large firm, 'You can call it a "good ole boy" network if you like, but I tell you these relationships motivate people. To solve delays we mobilize worldwide pressure.' Another Director of International Operations sounded less pugnacious:

> Success depends on building relationships with engineers, vendors, and many other people . . . a personal relationship of trust. One has to develop that kind of relation so that one can get on the telephone and solve problems quickly and without fuss. Reputation is what ultimately matters. Project integrity is number one. We can be construction managers of some Japanese company in one project and their partners in another. With a reputation no one asks you, how could you be managers here and partners there?

4 Business Strategy

American business strategies follow from the characteristics already discussed: high site costs, technological leadership in design, and organizational skill. Diversification is usually involved. Americans try to push high technology and better organization into areas previously dominated by others, either in competition or in association. The leading firms do not appear reluctant to displace or join other American or foreign enterprises in this pursuit of higher profits or flight from losses.

Preferred Types of Contract

What type of contract or responsibility companies prefer to have, given market conditions, is best shown by what they are actually doing. That preference has already been shown in Table 2.5, and diversity is what the table illustrates. Design-construct is the exclusive activity of some, while others mainly do construction management. In the case of Kellogg, the country's leading firm, all CM is carried out for plants of its own design. Despite the stress on design and management by many, other firms, especially smaller ones not shown in the table, remain general contractors or special subcontractors.

Not preferred is high risk, hence competitive bidding, especially overseas. Bidding on civil engineering projects and traditional building types, for which they lack a cost advantage, is not attractive to American firms if a foreign government is the sponsor or an international bank the financier. Foreign firms are thought to benefit more from subsidies through loans and tax relief.

Construction management contracts, like design-only contracts, are supposed to be awarded on the basis of qualifications, not price. Firms are prequalified, then short-listed, so that only four or five are actually competing with their presentations. A 'two-envelope' system is supposed to insure that the schedule of work proposed in the first envelope determines the choice, while the second envelope with the price is not opened until that choice has been made. But since all short-listed firms can presumably do the job well, price is usually expected to be the deciding factor after all, and many hints about it can be put in the first envelope.

Competition among a selected few also takes place for design-construct and construct-only contracts awarded by multinationals, foreign enterprises, and some governments. The price may be negotiated, a lump-sum, or it can be cost-reimbursable plus a fee. The fee itself can be negotiated and there can be a fixed maximum for the project. Austin claims to be the pioneer of design-build and fast-track working, establishing a rather firm price after preliminary designing and analysis and beginning construction before detailed drawings are complete. The company contracts in no other way, and says that five to fifteen per cent of time and money can be saved with this procedure.

Other firms are equally committed to a negotiated lump-sum. 'We do no such cost-plus contracting', said L. W. Donnelly, Senior Vice-President of Dravo. 'Ninety-nine per cent of our work abroad is EPC [Engineering, Procurement, Construction]. Of course, in our lump-sum EPC bids we put a value on our technology. All know that's how it's done by us. It's the secret of our success.' In late 1987 Dravo planned to sell its construction divisions to Jacobs Engineering and Westinghouse Electric. Being 'too aggressive in bidding' had 'battered . . . earnings' (*ENR*, 27 December 1987). Note that by taking on responsibility for the design, contractors may need massive indemnity insurance.

Bidding Tactics
Most details of bidding strategy are secrets of top management not shared with outsiders writing books or even with most of a firm's own employees. Nevertheless, senior vice-presidents directing international operations will usually reveal one or two details that together with tidbits from other firms suggest a general pattern for American firms. The pattern is intriguing but not greatly different from what one would expect (Clough, 1975; Lange and Mills, 1979; Hillebrandt, 1985).

First a decision is made that a project fits the company's annual bidding plan and the resources allocated to the estimating department. Then engineers, managers, estimators, and accountants work out the probable cost, including typical variations and worst-case scenarios. A report is taken to a small group of top executives who will raise sceptical questions and will ask for revisions and elaborations for a second, third, and fourth meeting. Eventually the small group will meet alone to decide on the lump-sum bid, the percentage fee, and whether to participate at all. The bid or fee will reflect the markup for profit, overhead, and technology although this markup will be partly hidden from clients, stockholders, and tax collectors through charges and billing rates that are higher than expenses. The actual markup reflects a firm's current objectives, involvement in other projects, and an assessment of the competition (their order books, expectations, and past behaviour).

A thoroughly prepared bid or proposal may cost as much as one-half or one per cent of the gross value of the work to be carried out, depending on the extent of novelty. Of course, the markup itself raises the gross revenue from a project if the contract is won. How many projects a firm can pursue depends on the cost of preparing the proposal compared with expected profits. If profits are in the range of 3 to 5 per cent, then firms must seek to win once in four or five attempts, or proposal and bid-making costs will exceed profits. Since American construction executives used such figures in the mid 1980s, they tended to avoid projects sought by more than four or five serious competitors.

The deviations from this basic pattern are many and reflect rising or falling business conditions and workloads. In good times some companies try to be one of only two or three competitors. In the peak year of 1980, the median level of profits as a percentage of revenues abroad was reported as 10.3, and the average for the upper quartile of American firms was 16.5 per cent. At such levels one can cut a few corners in making proposals, reducing their cost and at the same time bid on more projects. Some firms were willing to win one in ten. Entry is encouraged and prices as well as profits are bound to fall. As early as 1981 the median profits from American overseas contracting, mostly in the Middle East, had fallen to a mere 1 per cent, less than half that of firms from other countries (*ENR*, 16 July 1981 and 15 July 1982). As many as twenty firms competed for some projects, and American firms lost interest in several categories.

The presence of no more than four or five competitors does not, of course, imply that the chance of winning is necessarily 20 to 25 per cent. Some of the four or five may have the inside track. One company, for example, will no longer respond to invitations to bid in Taiwan if Bechtel or Ebasco have been invited: 'Why should we bother, just to be a reference point? Bechtel and Ebasco have long-established subsidiaries and contacts there.' Another company is reluctant to bid if two or more Japanese firms are competing. 'We've seen it time and again. One Japanese bids shamefully low but forfeits the bid bond if the higher Japanese is still below us. They let him get the contract. Maybe discouraging us is their game.'

In design-build competitions for process plants, firms offer different design and cost combinations, including both construction and operating costs. Each competitor has a different package with a different lump-sum price. The construction manager or private client can then play one bidder off against another and negotiate the price down. 'It's unethical, but it's done', said the international manager of a firm making highly specialized steel structures. 'Multinational clients are very hardnosed. We trained and licensed one Japanese firm with the understanding that they would not compete with us outside of Japan. Then one of our own multinationals set them up in the Middle East, to do just that, compete with us.'

Executives of another company reported that a competing German bid was revised downward from $100 to $75 million. 'We don't see how they could do it for that. You never get the details of a winning proposal, just that we were especially costly in this or that. But we suspect the German government gave them some tax write-off or other subsidy so they wouldn't have to pay unemployment compensation, saving money. If that company is involved in the future, we won't participate.' A third contractor for similar reasons withdrew from competing with Canadian firms to lay pipelines in Canada.

These assertions must be taken with a grain of salt. As an international banker in Washington said to us: 'Losing companies make non-stop allegations of tricky deals.' Still, there may be many tricky deals one never hears about.

Foreign Associates and Subsidiaries
Bidding strategy is influenced by the views of a firm's foreign associates. The effect will be minimal if the associate is a wholly owned subsidiary or a paper organization set up merely to comply with nationalist laws. More bargaining and exchange of information takes place when the foreign associate is a genuine partner or leader of a consortium with access to finance or with political clout. American firms abroad have all types of associates and subcontractors, and they may be subcontractors themselves. In that case their dealings are like those with multinationals as clients, discussed above.

Due to technology transfer and education abroad, the era of paper organizations has nearly passed. 'In Saudi Arabia', said one manager, 'We worked with a one-man firm, Abdullah Engineering Associates, or something like that. Nominally he was the CM, but as performance guarantors, we had to sign every document. Now the Saudis have firms sufficiently qualified to do the job by themselves.'

Wholly owned subsidiaries begin when a company finds it worthwhile legally and fiscally to incorporate a foreign marketing or management office. Without such foreign branches, contractors like Brown and Root or Stone and Webster would have found it impossible to continue in Libya when the US government forbade such work to Americans in 1986. As already mentioned above, major US subsidiaries in more advanced countries can do everything from feasibility studies on and employ no Americans. They merely have to check with their American parent, especially when contracts are being negotiated; and they must endure the occasional visit by a senior vice-president.

The number of subsidiaries abroad in 1985 for a few companies was as follows: Austin, 12; Dillingham, 6; Dravo, 12; Ebasco, 9; Fluor, 8; Kellogg, 9; Morrison-Knudsen, 3; Parsons, 3; Pitts-des Moines, 6; Stone & Webster, 9. Most companies have one subsidiary in Saudi Arabia and the rest about evenly divided between industrialized and developing countries. The most likely countries are Britain, Canada, Australia, Mexico, Argentina, Singapore, Hong Kong, and Indonesia. Some of these branches have their own subsidiaries. Anomalies of various types abound. For example, apart from building some aircraft repair facilities in Iran, the Austin Company has chosen not to be involved in the Middle East and lacks subsidiaries there. It has sold the Italian subsidiary to its former manager and allows him to continue to use Austin's logo and technical support. It has had joint ventures only with other American firms. In addition to having *pro-forma* host associates and overseas subsidiaries, therefore, American firms will join with each other or with whatever foreign firms, public or private, might be helpful in winning a contract. In the Far East a typical combination is American design and construction management, Japanese finance and procurement of materials, and Korean sitework.

Companies that specialize in traditional civil engineering are likely to have fewer foreign joint venture partners, perhaps two or three. Thus Morrison Knudsen has joint ventures with Germans and Danes; Perini with French, Italians, and Canadians; and Atkinson with Venezuelans, Chileans, and Brazilians.

There is no prejudice against working with public sector partners. 'On the contrary we like a captive contract like that', said one director for international operations, 'It leads to tremendous gains.' In a Mediterranean country that company was partners with both local contractors and the client public agency. Over a period of twenty years the company's local subsidiary had grown to over 2,500 employees, including only one American. In the Far East the company had won a similar lucrative contract for building roads and bridges as a joint venture with a government company.

5   Relations with Government

Dealings of international construction contractors with the American government can be described by one word: fragmented. Neither firms nor government have a single coherent policy toward influencing one another and promoting American interests at home or abroad. Among the firms, designers, builders, and designer-builders have separate lobbying organizations often with different objectives and strategies. Separate government agencies are charged with promoting the industry's interests, and these activities are only haphazardly checked by an Inter-Agency Major Projects Co-ordinating Group consisting of the US Trade Representative, the Export Import Bank, the Agency for International Development, and the Departments of Commerce, Labor and Treasury. Some relevant agencies can be led by persons who either ignore the industry's interests or question the soundness of any government support. Moreover, the US Congress may pass legislation that puts these agencies at cross purposes and that may be harmful to the sector's foreign or other activities. In the words of one industry spokesman,

> Everything in Washington is inefficient and screwed up. Every agency wants a finger in the foreign trade pie, and we wind up speaking with six voices and get no effective policy in a timely manner. You should hear the vituperative statements that one agency makes about another when they ought to be on the same team. Everything is terribly nonproductive and wasteful.

Ideology and Lobbying
Above all, one should not believe that the presence of former construction executives in leading government positions, even as Secretary of State and Secretary of Defense, among others, gives the industry a better hearing at the White House, especially in a Republican administration. A coherent government–military–industrial complex for exploiting the rest of the world is not their outlook or ideology. They do not believe in 'anything for business'. Right or wrong, these former (and future) businessmen in government believe in sink-or-swim market place individualism. Heading a government agency to them is simply heading a different sort of enterprise that has adversarial relations with other enterprises. Their instincts are to be competitive, to bargain, to make deals, yes; but from the perspective of their agency, not the community, not even the business community. Economic advisors hired by such men are likely to be free traders who believe that any type of export support, even retaliatory, is as bad as any type of tariff.

Officials of the different contractors' and designers' associations tend to each have their own vision of co-ordinated US government support for their members. Four of these groups in 1967 associated themselves in the International Engineering and Construction Industries Council (IECIC). These four are the Association of General Contractors of America, the National Constructors Association, the American Consulting Engineers Council, and the American Institute of Architects. But 22 others such as the Associated Builders and Contractors, the National Association of Home Builders, and the Construction Management Association of America are not represented in IECIC. All 26 are part of a National Construction Industry Council that in turn has set up a foundation, the National Construction Industry Research and Education Foundation, for studying the (detrimental) effects of taxes on the industry. This study, itself tax-exempt, was then subcontracted to Data Resources, Inc., the econometric consulting firm.

The higher the level of government, the less sympathetically are the industry's views received. Indeed, top officials of the Reagan administration tried hard (and nearly succeeded) to abolish the Center for Building Technology of the US Bureau of Standards, the Foreign Commercial Service, the Trade and Development Program, and the United States Export–Import Bank, the government's primary agencies for construction research and for supporting exports of construction services. At lower levels some Federal Departments have sections specialized in construction with officials who identify rather fervently with the industry's problems. Their reports and recommendations tend to have a partisan tone. Higher up, for example with the US Trade Representative (who negotiates on tariffs and trade barriers) such reports are treated with scepticism. If they nevertheless lead to treaties or proposals for legislation, committees of the US Congress are likely to re-examine issues, for example, through the Congressional Technology Assessment Office.

To encourage Congressional support, the various construction industry associations have 'political action committees' (PACs) that begin each election year with fundraising programmes among construction industry executives. According to law, contributions to political candidates must come out of the pockets of individuals, not from corporations. Nevertheless, PACs can accept corporate cheques for their administrative expenses, including those of fundraising, and apply all personal contributions to actual political campaigns. Funds involved are comparatively modest. In 1986 the National Constructors Association, representing the industry's largest firms had a fundraising goal of only '$10,000–$15,000 per election cycle to allow participation in at least a dozen House races and six or more Senate contests . . . to provide financial support to candidates . . . who have been identified as supporting the goals and interests of the association' (*NCA Newsletter*, January/February 1986, p. 3). Corporations may, however, contribute as well to association funds that pay members of Congress for speaking at industry meetings. According to the chairman of one association, 'political fundraising is the cornerstone of government relations'.

Among Congressmen who have proposed laws urgently sought by the industry are Don Bonker, Democrat of Washington; Nancy Johnson, Republican of Connecticut; Dan Mica, Democrat of Florida; and Senator Frank Murkowski, Republican of Washington.

Policy Goals
What the industry seeks from Congress is more discrimination in its own favour for carrying out US-funded work abroad, help with identifying projects funded by foreigners, removal of any hindrances to

winning contracts abroad, pressure to open foreign markets, and access to finance on favourable terms. The industry is uncertain about needing government assistance with technological research and about seeking protection against foreign contractors bidding for work in the United States.

American hindrances to working abroad are mainly the 'anti-boycott' legislation of 1976, the Foreign Corrupt Practices Act of 1977, and taxes on American overseas work. The anti-boycott legislation forbids US firms from seeking contracts by complying with Islamic bans against dealing with Israel. It is not boycotting *per se* that is condemned, but boycotting of Israel. The US Congress has declared any firm *not* boycotting Libya as ineligible for bidding on embassy construction projects. One Texas firm reported that 300 Americans lost their jobs because Libyan work had to be rearranged.

The Foreign Corrupt Practices Act puts heavy penalties on attempts to bribe foreign officials in order to obtain contracts. Criminal convictions against company presidents are possible although none had been obtained by 1986. The industry does not lobby for permission to bribe, even though bribes may be customary, though perhaps illegal, in many host countries and although they may be viewed as legitimate tax-deductible business expenses in some European countries. The persistent complaint has been that the language of the Act is so vague but menacing that much business, especially in joint ventures, has been lost. The House of Representatives, but not the Senate has passed an amendment allowing 'irregular payments' to foreign 'clerical' but not 'ministerial' decision-making officials. If bribes are paid to ministers by foreign intermediaries that a company must employ, the standards for expecting the company to know about it remain unclarified by Justice Department rulings.

Year after year industry representatives testify at Congressional hearings about anti-boycott and corrupt practices legislation, yet they do not necessarily want the issues fully reopened for fear that Congress may make things worse. As one corporate Director of International Operations told us,

> The Corrupt Practices Act has all sorts of constraints about who can be an agent and how they can get paid. They can't be a government employee, for example. So we don't hire those, but we have agents. You need them. Same thing applies to anti-boycott legislation. The way it is now, it has language that both we and the client can live with. Anyone can. When it comes to real corruption, we simply avoid countries that are known for that.

Taxes
The most annoying hindrance that Congress is asked to remove is taxes. Unlike nationals of other countries, Americans working abroad have to pay foreign and US Federal taxes, raising the cost of employing them. At the same time some countries tax preliminary work carried out in the US and also taxed there.

After 1981 American employees abroad were allowed an exemption of $80,000, but allowances for housing and schooling were henceforth to be counted as income. In the tax reform bill of 1986, after attempts to eliminate it altogether, the exemption was lowered to $70,000. The president of one major contracting company confirmed to us that, 'Limitation on the US overseas salary exclusion policy would have a serious impact on our ability to provide a competitive proposal.' Another executive said that elimination of personal taxes for Americans working abroad would be 'the most beneficial policy change' because 'US companies are priced out of the market due to taxation that our foreign competitors do not have.'

However, US salaries, wages, and fringe benefits earned abroad have averaged only 4 to 5 per cent of contract value (Price Waterhouse, 1985, pp. B–2, B–3), and for some major firms they were even less than 1 per cent of overseas wages and salary costs alone. It therefore appears doubtful that any increase in bids to cover extra taxes could have been that critical in losing contracts. Spokesmen for two of the largest and most successful firms confirmed that view. What taxes damage most are the personal incomes of one's friends and colleagues.

The US Government as an Overseas Client
As operator of embassies and military bases all over the world, the US government is itself a major client for international contractors. Almost nothing kindles wrath among American contractors more than losing taxpayer-financed projects to foreigners. They like having Indian Ocean work reserved exclusively for themselves, and they are not unhappy about a 20 per cent bid preference in their favour in the Pacific.

In general the US government is considered a safe but peculiar client. Making a presentation to the Navy has been compared to a court martial. The Army is said to be more informal, more inquisitive, but less predictable (Lucas 1986, pp. 102–3, 121–2). Most criticized is the State Department's Foreign Building Office, which is currently managing a $2.9 billion programme to make US embassies safer. A Congressional committee recently accused this Office of 'serious and longstanding systemic deficiencies . . . substandard design, poor contracting procedures, shoddy workmanship, inadequate management and major cost overruns' (*Washington Post*, 20 August 1986).

American contractors have claimed that one of their firms as programme or construction manager could have prevented all that confusion. Their vigilance focuses on reserving any embassy work in excess of $5 million for American firms via a 10 per cent bid preference. That the Cairo embassy is being constructed by the American subsidiary of the Japanese Kajima has not pleased them. New legislation introduced by Congressman Mica is designed to limit embassy construction to firms who will subcontract less than half of the work, employ Americans in 80 per cent of supervisory positions on the site, have sufficient technical and financial resources in the United States to perform the contract, and have more than five years of legal existence in the US. In January 1986 a Japanese mission came to the US to complain about these and other American preference policies (*International Construction Newsletter*, January 1986, p. 3).

Trade Promotion
Bidding low safely depends on information, and all contractors stress that learning about a project early, in the 'gleam-in-the-eye' stage, is crucial. To render that service, the Carter Administration set up the Foreign Commercial Service in the International Trade Administration of the Commerce Department. Contractors generally praise this service but usually add that it could be faster and bigger and is still not as good as the British equivalent. In response the Department has taken such measures as placing a part-time procurement advisor in the office of the US Executive Director of the World Bank. Her job is to facilitate US access to World Bank contracts through early information.

The Office of International Major Projects of the International Trade Administration also conducts such export promotion activities as organizing trade fairs and trade missions. Contractors are especially fond of being on trade missions because, as one testified before a Congressional Committee, 'personal contact is the primary, the best, and perhaps the only way to successfully sell services overseas' (J. K. Morrison, Testimony before the House Foreign Affairs Subcommittee on International Economic Policy, 24 May 1983). The speaker went on to say that follow-up after the missions by embassies could be better and that the ambassadors themselves should get involved, as he believed the British and Japanese do. However, one type of follow-up that is lacking is due to the contractors on the missions themselves. They are reluctant to report in detail on just what they accomplished on the grounds that such proprietary information about bidding and procurement tactics could fall into competitors' hands through the Freedom of Information Act. Hence, such reports are no longer requested.

Lowering Trade Barriers
Most countries protect their construction industries, not through tariffs, but with other preferences and discriminatory measures. To capture access to foreign markets, governments first have to identify these trade barriers and then negotiate lowering or removal. Negotiation may be bilateral, as currently with Canada and Japan, or multilateral under the General Agreement on Trade and Tariffs (GATT). The United States Trade Representative (USTR) is in charge of these proceedings.

Usually negotiations take place in a wider context involving activities besides construction. In 1986 a bilateral example was the plan for a free trade agreement between Canada and the United States. In preparation, the principal construction trade associations through IECIC, described above, were developing a position paper. In this case an American objective was modification of 3–15 per cent bid advantages on provincial public sector projects given to Canadian firms with head offices in the particular province giving the advantage.

More discordant were American charges of arbitrary exclusion from bidding on the $8 billion Kansai airport planned for an artificial island in Osaka Bay, Japan. Against the predictable opposition of the Japanese trade associations, Prime Minister Nakasone had promised that bidding would be open to all on public projects. Yet despite more than 80 per cent public ownership, the Kansai project was classified as private. More than 20 American companies registered for airport work, although the announcement seeking bids had been no more than 'a small, hard-to-find advertisement in the back pages of a Japanese-language newspaper' (*New York Times*, 1 September 1986). 'We will continue to apply pressure through all available channels at each and every opportunity', testified a Commerce Department official before a Senatorial subcommittee (*Washington Post*, 6 June 1986), and by November 1987, $2.2 million in Kansai contracts had gone to American firms, a minute portion of the $2 billion so far awarded. An American negotiating team failed to win further concessions from the Japanese with respect to an expected $62 billion in public works contracts before the year 2000, and as a result Congress banned Japanese construction firms and their US subsidiaries from federally funded projects in fiscal 1988 (*ENR*, 10 December 1987).

Bilateral solutions to problems with trade in services are difficult because no internationally accepted code of behaviour exists. The USTR, Clayton Yeutter, said that developing trade rules that will open markets within the framework of GATT had high priority for the United States (USTR, *US National Study on Trade in Services*, December 1983). But in 1986 the American construction industry had 'not yet reached a consensus on whether or not the potential benefits of a trade negotiation on services would outweigh the possible costs to our industry' (IECIC, 'Paper on Gatt', 1986). They were afraid that their interests might be traded off against those of other service industries, giving benefits to those other service exporters and losses to construction at home and abroad.

An international code would eliminate foreign requirements that engineers be graduates of national universities and that companies should be locally owned. Various other restrictions on employment and corporate licensing would cease to be discriminatory. US companies would not have to make more financial disclosures than national ones. 'Buy national' requirements for government procurement would be eliminated. Since American firms benefit from such provisions in the US market and in working for their government abroad, they have to compare gains with losses at home.

Once again, firms seem especially apprehensive about agreements that might make them divulge information about where and at what cost labour and materials are being procured. They fear that contract awards will be challenged by competitors just to smoke out bidding tactics and management techniques. Such challenges were seen as especially threatening when awards are based not on price but on 'state-of-the-art fast-track project management capabilities', the very thing Americans do best (IECIC, 1986, p. 7).

Help with Finance

Finance of construction is sufficiently complex so that public funds can be brought in through a variety of disguises and channels. Basically, technology and productivity may be more important, but as John C. Richards, Vice-President of the M. W. Kellogg Company, testified before a Congressional Committee: 'No single element is more crucial to . . . our industry as a whole than the availability of adequate, reliable, and competitive export financing' (House Subcommittee on International Economic Policy, 24 May 1983).

In our survey of firms, the same priority was affirmed time and again. Said the Director of International Operations of an East Coast firm: 'What heads the list of problems for an American firm is that we don't have enough of a lending agency that helps on a competitive basis.' The President of another Eastern firm said: 'Project finance is frequently as important as pricing. Many countries have project needs that are totally dependent on the availability of financing.' An executive of a large Western firm said: 'More than ninety per cent of [our] international work over the last ten years has required project financing assistance of one sort or another.'

In response to such pressures, some contractors have tried to work jointly with private banks, and others have hired bankers and set up their own internal financial divisions. One of the earliest was Bechtel Financing Services, Inc., which, established in 1969, moved gradually from export finance to comprehensive project financing. In our survey we found that firms above an approximate billion dollar threshold in foreign contracts were willing and able to tap private capital markets in this manner, but that smaller firms were not. This difference may partly explain why the top firms greatly expanded their international market share during the 1980s, as shown in Section 2 above. With their overseas subsidiaries, large firms like Kellogg, Fluor, and Bechtel have also been able to arrange financing in European capital markets, sometimes at foreign subsidized rates.

When private funds cannot be borrowed at low enough interest rates, then a good alternative is government money. Contractors and other exporters persuade Congress to channel funds their way by complaining about the cunning but generous treatment of foreign competitors by *their* governments. Thus the Overseas Private Investment Corporation, OPIC, was set up as a government agency under the Carter administration to provide risk insurance against currency inconvertibility, confiscation, war damages, and a variety of disputes. That brings down risks, hence interest rates. The Trade and Development Program of the State Department pays for prefeasibility and feasibility studies. Most important, however, is the Export–Import Bank of the United States (Ex–Im Bank).

The Ex–Im Bank

The Ex–Im Bank was established in 1934 to finance export sales of American goods and services. By the mid-1980s it had 15 major programmes for insurance, guarantees and lending. Most important for contractors is its new ability to lend at subsidized interest rates for up to 85 per cent of the costs of a project for both exported services and equipment. Formerly the share that could be covered was 65 per cent. Mainly because of low interest rates, the bank lost about $380 million in 1985.

The Reagan administration twice wanted to abolish the Ex–Im Bank altogether as a money loser and needless competition for private banks. By the autumn of 1985, however, enough contracts had been lost, and the American balance of payments had deteriorated sufficiently to cause a reversal of government policy. In five years the use of mixed credits by other countries had more than tripled to $9 billion. A $300 million 'war chest' was requested from Congress. William H. Draper III, President of the Ex–Im Bank, said, 'We want to see what damage we can do to the French and to others [using mixed credits]. . . . We want to bring about a successful peace in this war of mixed credits' (*Washington Post*, 13 November 1985). France was not the only country offering mixed credits but was targeted because of 'obstructing negotiations'.

In anticipation of this fund, the Ex–Im Bank lent money to help Scientific Atlanta win against French and Japanese competition in early 1986 for building 12 earth satellite stations in Gabon for $20 million. Congress, however, delayed approval of this fund, which could be used for all exports, not just construction services. The administration meanwhile sought to finance the programme with funds diverted from the Ex–Im Bank's direct loan programme. This programme had itself been reduced from several billion to $900 million. The industry felt that not reducing the loan programme further was more in its interest than the mixed credit programme.

Efficiency, Finance, and Foreign Aid

Conservative Republicans were happy to see the mixed credits idea founder. Lawrence A. Kudlow, who had been with the Office of Management and the Budget during Reagan's first term, said:

> There is no evidence that [export credit subsidies] help win contracts. It's a waste of resources to subsidize and protect companies. They should be disciplined by the market place. What we're seeing now will only lead to greater inefficiencies, flawed bidding and poor products. (*New York Times*, 2 February 1986)

The industry's point of view was quite different. As one Senior Vice-President told us:

> International work means beating the Japanese. But how can we do that? The US government and American financial institutions haven't been partners with us, the way the French and the Japanese do it.

Many contractors seem to share this view that foreign aid funds should help them get contracts in Third-World countries. The targeting of American foreign aid toward agriculture, health, and education for the poor majority strikes contractors as misguided and arrogant; but it has remained a Congressional mandate except for military assistance and the related Economic Support Fund (ESF) for 'front-line' states and countries that allow 'military access'. Construction lobbyists think ESF funds should be used for mixed credits for their projects and clients. 'We have fought them tooth and nail on this', said one industry leader.

Conceivably, AID might have a bigger constituency and amount to a larger percentage of national product if it catered more to interest groups like the construction industry. In 1984 the American official development assistance share of GNP was 0.23 per cent, compared with 0.77 per cent for France.

Support for Research
Advised by its Technology Assessment Office, Congress may eventually conclude that American contractors are in danger of losing business abroad, not only because of unfair government–business co-ordination in competing countries ('lack of a level playing field'), but because of inefficiency, lagging research and development.

Indeed, American technological backwardness on the construction site (by contrast with the management and the design of the structures being built) has become widely acknowledged. During the 1970s construction productivity declined sharply in the United States, raising costs to a point that led to a large private Construction Industry Cost Effectiveness study, headed by the Vice-President for Engineering of the E. I. du Pont de Nemours Company. The conclusion was that lack of data and research, by contrast with other countries, was a large part of the problem.

As a result of the study, 31 contractors, 30 major clients, and 22 universities joined forces in 1983 to set up a Construction Industry Institute (CII) at the University of Texas with an annual budget of about $2 million. So far CII has concentrated on American problems, noting that sites are not chosen with sufficient concern for accessibility and that constructability is still being slighted by designers. CII works through task forces on problems of productivity measurement, contract types, cost controls, materials management, quality management, employee effectiveness, safety, and others. Apart from some work on induction pipe bending and fibre-optics, it has not stressed technology as such.

Other American construction research goes on at state highway departments and under the auspices of other large buyers of facilities. For example, electric power companies have set up the Electric Power Research Institute. The National Science Foundation has programmes for studying earthquake mitigation, for analysing the performance of existing large structures, and for novelties in building systems and automated construction. Centres have been set up at Lehigh and Carnegie Mellon Universities. The military have six construction research laboratories and have given $14.5 million for a Center for Advanced Construction Technology at the Massachusetts Institute of Technology.

As a whole, being fragmented, the industry (with the major exception of Bechtel's laboratory employing about 300 people) has shown little interest in financing research itself through institutes, universities, or government agencies. Amounts expended are unknown but they are far below the $120 million spent by Japanese firms in 1980. As a percentage of sales, the Japanese construction industry was reported to have spent about 3 per cent on research in 1984, while the US industry spent about 0.01 per cent (*ENR*, 31 July 1986). The Center for Building Technology of the National Bureau of Standards, the largest American construction research centre, was nearly abolished, as mentioned above, and still has a budget of only $12 million. It operates under the suspicion that government is trying to compete with business, anathema to American conservatives.

The latest attempt to study and bolster the competitiveness of American contractors abroad is being made under the auspices of the Building Research Board of the American Academy of Science and Engineering in Washington. Their project on international construction initially grew out of concern at the academically oriented National Science Foundation for better seismic construction methods. Whether major support by American contractors can be mobilized remains to be seen. Construction firms in general have behaved as if materials suppliers or equipment makers should finance and develop new construction methods. Their own job is organizing factors of production to be 'on schedule and within budget'.

6 Conclusion

Officials and lobbyists in Washington who seek benefits for the industry are likely to make any clouded outlook seem like an impending storm. One hears that, 'Yes, the outlook is grim. Our number of firms among the top 250 is down, the Japanese are up, and so is their dollar volume. Same with the French.'

Many of the foreign gains have come in the United States itself where US contract awards are not counted as 'international'. From 1981 to 1985 awards to foreign international contractors *outside* of the US fell by 55.2 per cent from $90.8 billion to $40.7 billion (1982 dollars). In the same market the share of US contractors also fell drastically from $47.4 to $25.3 billion, but *proportionately* somewhat less, 46.7 per cent. The American share rose from 34.3 per cent to 38.3 per cent. What happened to US overseas contractors in general was therefore not 'grim' beyond the general collapse of the market. Severe effects were felt by firms in energy-related construction types, especially those building American nuclear power plants with their controversial record. High costs and lagging technology did not help American firms in conventional building and infrastructure. These fought hard for a share in

American-funded overseas projects and were extremely sensitive to real and imagined collusion among foreigners and their governments. But even the most advanced design-construct firms building chemical plants, data processing centres, satellite tracking stations, and the like, did not feel very secure in their niches.

TAPAN 4420
4210
6340

3

Japan

YASUYUKI HIPPO and SABURO TAMURA

1 Introduction: the Development of Japanese Overseas Construction

Japanese overseas construction activity began at the turn of the century with a railway between Seoul and Inchon in Korea. For the next 60 years it was dominated by colonial, government and military works. Only after the completion of postwar reparations in 1966 did construction exports begin to take off on a commercial basis. From 1973 onwards, overseas construction contracts expanded rapidly with peaks in 1979 (boom in Middle East) and in 1983 (boom in South-East Asia). The value of contracts received in 1983 topped 1 trillion yen (4.4 billion US dollars) for the first time. By 1984 Japan had become the second largest exporter of construction services, following the USA, as well as the second largest donor of official development aid (ODA) overseas. Despite rapid increases in recent years, overseas activity still accounts for a relatively small proportion of total Japanese construction output. As is well known, the Japanese construction industry is huge, employing 5.4 million people, or 10 per cent of the total labour force in 1986; domestic construction investment generally contributed approximately 20 per cent (gross) to GDP in the 1970s, and more than 15 per cent even in recent stagnant years. However, in 1985 overseas orders accounted for only 2.3 per cent of national construction investment, and 2.5 per cent of Japan's total exports on a goods basis and 9.8 per cent on a service basis. The top 50 firms had overseas orders of 9.2 per cent on the average.

This situation is now expected to change, however. Because of stagnation in investment in public works (reversed during 1986–7) and sluggish demand from the private sector, positive promotion of 'Construction Exports' has recently become a part of the 'tri-directional business strategy' of general contractors, along with

'Diversification into Numerous Fields' and 'review of corporate management by the use of the Total Quality Control concept and technique'. Before looking at such strategies in greater depth in section 2, a brief outline of the history and development of Japanese overseas construction would seem to be in order.

Overseas Construction before the Second World War

From the turn of the century to the beginning of the Pacific War (1941–5) – i.e., the period of Japanese colonial expansion – a series of railway works by native labour forces dominated the Japanese overseas construction scene. After the Seoul–Inchon railway (1897–1900), a ten-year plan for the construction of a 'Trans-Formosa Railway' from Jilong to Gaoxing (a distance of 400 kilometres) was implemented. Then, in preparation for the Russo-Japanese War of 1904–5, the Pusan–Seoul and Seoul–Sinuiju lines were completed. Contractors on these projects suffered financial losses and even bankruptcies due to inadequate and inaccurate information provided by the client. After the war, the South Manchuria Railway Company was established, and the Shenyang–Dalian line opened. Thus the two outlying cities of Pusan and Dalian were finally connected by a colonial railway.

The establishment of the 'State of Manchuria' in 1931 enabled Japanese contractors to rapidly expand their workload in the colonial zone in diversified construction fields including railways, new hydroelectric power stations and mining development. By the opening of the Pacific War in 1941, Japan's overseas construction activities stretched from the north to the south of China, and on to South-East Asia where military works (roads and airports) were of vital importance. These early overseas construction experiences differed from modern construction exports in two ways: first, prewar overseas construction sites were either in Japanese colonies or in areas under military occupation by Japan; secondly, the owners or clients of the projects were Japan's army and navy, colonial government, or special companies founded as a deliberate act of national policy. Hence, payment was not in foreign currency. Only one contract was carried out in an independent foreign country on a strictly commercial basis: in 1937 member firms of 'Kyoei-kai' (an overseas consortium formed by Shimizu, Ohbayashi, Takenaka and Hazama) undertook to build a road between Veracruz and Jalapa, a distance of 50 kilometres, for the State of Mexico. The project was completed in seven months. Kyoei-kai eventually became a prototype for OCAJI (Overseas Construction Association of Japan, Incorporated), formed in 1955.

The Period of US Military Works in Okinawa (1945–54)

As a result of Japan's defeat on 15 August 1945, overseas military works came to an end; but in their place, the US Occupation Army's construction boom broke out in the Japanese Archipelago. This nationwide boom reached its peak in 1947, to be followed by a period of over-competition. To reconstruct Japan's construction industry which was in disorder, the Ministry of Construction was founded in 1948, and the Construction Industry Law incorporating a system for the registration of contractors was passed in 1949.

In 1950, Japanese participation in US military construction works in Okinawa was permitted for the first time. Following the outbreak of the Korean War in June 1950, Japanese contractors rushed to the Okinawa works; 25 firms were engaged in the peak year of 1952. As Japanese contractors had made little progress, either in technology or in business style since the Sino–Japanese War of 1937, their contact with the American way of working at Okinawa helped them to master US large-scale mechanized techniques, contract system, and site-management style. This experience was developed further in domestic base-works at Atsugi, Misawa, etc.

In the Okinawa work a notable event that took place in 1950 was the development of joint-venture projects between three Japanese contractors (Kajima, Takenaka, Ohbayashi) and four US contractors (Morrison-Knudsen, the Bechtel Group, Pomeloy, Kiewit). This was the first experience of its kind in the history of the Japanese construction industry.

The Period of Wartime Reparation Works (1954–66)
After the peace treaty with the USA in 1952, Japan became an independent country. Along with a domestic boom in hydroelectric power development, wartime reparation agreements were concluded one after another between Japan and South-East Asian countries. The first overseas construction work undertaken as part of a reparation agreement was the Barûchan Water-Power Station No. 2 in Burma, which was begun in 1954. From that time through to the Indonesian Kalankates Kalikonto dam works in 1966, reparation works continued, reaching a peak in 1962. Typical projects included tunnels and hotels in Indonesia, the Danim dam in Vietnam, and urban development in Hong Kong. Such projects, awarded exclusively to Japanese firms, served as a lever, enabling Japanese contractors to make inroads into foreign markets. During this period two new organizations began preparing for future overseas works on a commercial basis. One was OCAJI formed by major contractors. Another was IFAWPCA (International Federation of Asian & Western Pacific Contractors Association) founded by contractors from eight countries.

In the 1950s and the first half of the 1960s, there were various repair projects, the largest of which was said to be the Malaysian Muda River

Development (1960). But Japan's major contractors were then enjoying the expanding domestic market produced by high economic growth (as well as the construction boom generated by the Tokyo Olympics in 1964) and paid little attention to overseas works on a commercial basis. Consequently almost every overseas contract undertaken on a non-reparation basis suffered a loss.

Take-Off of Works on a Commercial Basis, 1966–73
It was not until the second half of the 1960s that overseas construction on a commercial basis really began to take off. Riding on the wave of the government's export promotion policy, overseas construction (incorporating machinery and materials export) was strongly supported. It was regarded also as a priority industry in contributing to the infrastructural sector of developing countries. Consequently, the Ministry of Construction established a new Japanese-style tax exemption system whereby deductions could be made from income earned from overseas construction work; as well as lower rates of bond-insurance, and the provision of finance by Japan's Export–Import Bank and Overseas Economic Co-operation Funds (OECF). Japanese overseas construction expanded, in line with Official Development Aid (ODA), until, in 1969, it exceeded $100 million of orders received from abroad. After the stagflation following the *Nixon Shock* (1971–2), the brilliant 'Overseas Boom Decade' opened in 1973, and foreign contracts jumped three-fold compared with 1972. The size of overseas construction jobs also increased enormously, as instanced by Kajima's construction of a one-million ton dock in Taiwan, Hazama-Gumi's Temengor Dam, and certain Penta-Ocean construction contracts of Malaysian and Singaporean harbour works exceeding 10 billion yen (approximately $33 million), each. Overseas construction orders were estimated to be 200 billion yen (approximately $666 million), in the fiscal year 1974.

As they made inroads into foreign markets, major contractors set up foreign subsidiaries: first Takenaka in San Francisco (1960), then Nishimatsu in Thailand (1963), Kajima in Los Angeles (1964), Fujita in Peru (1965) and Aoki in Brazil (1967). Since 1971 this movement began to accelerate in South-East Asia, the USA, and Brazil. By 1976 more than 74 subsidiaries existed in 21 countries including the Middle East.

From the Middle East Boom to Asian Markets
Japan's construction industry crawled out of the prolonged business slump started by the first oil crisis in two ways: first by heavy investment in the domestic public sector during 1976–9; and secondly by a remarkable rush into the Middle East, chiefly to Iraq, until the outbreak of Iran–Iraq fighting in 1980.

Petrodollar-rich Middle East nations planned big national development projects, including heavy industrial plants and equipment, and
infrastructural facilities; and they awarded contracts by international
tender which presented an unexpected business opportunity for
Japanese contractors. But Japanese construction concerns, although
trained for over ten years in traditional Asian markets, were
newcomers to the Middle East (with one or two notable exceptions,
such as Penta Ocean Construction and Toa Harbour Works). This
created enormous difficulties. Basically Japanese firms had to carve
out a market for themselves in competition with leading Western firms
on the one hand (with their rich experience in overseas projects aided
by excellent consultant firms) and on the other hand with Korean
companies that could offer cheap, qualified manpower and were supported by the Korean Overseas Construction Corporation (KOCC). In
order to cope with Western and Korean competitive power in international contracting, as well as with possible political risks, it was
generally acknowledged that Japan's system of government support,
in the form of special guarantees or loans to contractors, would have
to be greatly improved.

Despite this handicap, foreign orders almost doubled annually after
1972, resulting in a 7.5-fold increase between 1972 and 1976. Sluggish
performance in 1977 is largely attributable to Korean competition and
to the steep increase in the yen's value against the dollar. However,
the upward trend in orders resumed in 1978–9.

In 1976, 48.0 per cent of overseas contract awards were in the
Middle East; 38.3 per cent in South-East Asia; 3.1 per cent in Africa,
and only 1.0 per cent in Europe and the USA. By 1979 the Middle East
accounted for 55.0 per cent of all new contract awards and South-East
Asia for only 32.0 per cent. The outbreak of the Iran–Iraq war in
September 1980, dealt a heavy blow to a number of Japanese construction companies engaged in projects in the two warring countries.
In 1980 projects in Iraq included urban development in Baghdad, promoted by Shimizu Construction Co.; the construction of hotels,
hospitals and other buildings by Taisei Corp.; expressway construction by the Fujita Corp.; pier construction by the Toa Harbour Works
Co.; and a school construction project by the Mitsubishi Construction
Co. Due to the eruption of the war, the Toa Harbour Works Co. was
forced to suspend work altogether and to send its on-site workers,
including foreigners, back to their respective countries. Shimizu
Construction Co. reduced the volume of work to some 30 per cent of
normal levels. Japanese companies called on their Iraqi clients to pay
compensation, including the costs of evacuation and of depreciation
of machines and equipment left at construction sites. However, the
negotiations were hard and some companies suffered huge losses.

In the light of these events, the Japanese construction industry reconsidered its business strategy. This led to the pursuit of business opportunities in its traditional market of South-East Asia. Such large scale projects as the Hong Kong subway (1st stage 1973, 2nd 1978, 3rd 1982); the Asahan Aluminum Facility Contract in Indonesia (1979); and the Changi Airport in Singapore, had already started. In the early years of the 1980s, Japan's overseas contracts rose again from $2.1 billion in 1980 to $3.5 billion in 1981, $3.7 billion in 1982 and $4.2 billion in 1983. This upward trend can be explained chiefly by expansion in the Asian market, particularly in Hong Kong (subway, etc.), Malaysia (highway, etc.) and Singapore (dock works reclamation, etc.), as shown in Table 3.1. By 1982 the South-East Asian market accounted for 78 per cent of all new overseas orders; orders in the Middle East had fallen to only 12 per cent of the total.

Internationalization for the Pacific Age
Overseas construction contract awards in South-East Asia were at their peak (as a percentage of the total) in 1982. From 1983 on they began to decline due to the depressed state of the Asian economy. But this decline was paralleled by a steady increase in orders from the industrialized countries. With increased factory/office building by Japan's multinational companies in America, the USA was the country to place the largest number of new orders in 1984. In 1985 – when the total of foreign orders again reached the level of 1983 – the USA was closely followed by Australia (see Table 3.1). The regional concentration of Japanese overseas construction changed from the Middle East, to South-East Asia, to North America and the Pacific. For the time being, depressed oil prices and the difficulties of financing in developing countries, will probably insure that the relative importance of the US–Oceanic market will continue to increase. The 'Pacific Age' is coming to Japan's construction industry.

Deployment in the mature market of the US and Australia is increasingly characterized by a new type of urban development project. A typical example might be that of a Japanese contractor purchasing derelict buildings and designing new shopping centres or office buildings. This means that Japanese contracting firms are supplying not only high-level construction technology, but also services such as the provision of finance, co-ordination between those involved in the project, and regional development planning. In order to undertake this kind of project, contractors generally form a syndicate with financial organizations and real-estate firms. The actual site work, however, is left to local contractors – in order to avoid antagonism and accusations of market-snatching by Japan.

In this connection it should be pointed out that, although Japanese

Table 3.1　*Volume of Orders Placed with Japanese Construction Firms in Fifteen Countries*
(Yen, 100 million, selected years)

Country	1969	1977	1979	1983	1985
USA	9	—	—	780	1,988
Australia	—	—	—	540	1,763
Iraq	242	206	2,362	—	—
Iran	—	601	—	—	—
Saudi Arabia	31	359	561	438	—
Egypt	—	203	—	—	655
Libya	—	—	181	—	—
Hong Kong	61	234	391	795	1,058
Singapore	30	191	214	2,643	865
Malaysia	21	—	227	1,625	390
Philippines	12	156	—	—	—
Indonesia	33	211	190	1,063	390
Thailand	—	468	—	—	—
China	—	—	—	—	610
Sri Lanka	9	—	—	215	—

Source: Ministry of Construction, Japan.
Note: US $1.00 = 250 yen.

contracts in the USA (including Hawaii and Guam) today amount to over a billion dollars, there has been little penetration of the Japanese construction market by foreign contractors. In fact, there were only a very few recorded instances of major public-sector construction projects being awarded to foreign firms: one, in 1955, jointly pursued by G. F. Atkinson of the USA and Hazama Gumi at Sakuma Dam works, and another where in 1967, Potashenic Co. Ltd. of the USA won the international tender (occasioned as a result of the provision of finance by the World Bank) for the Meishin (Nagoya-Kōbe) highway construction work. In the latter case, however, the US firm subsequently dropped out of the project, due to their inability to cope with Japanese business practices and the multi-layered subcontracting system. This market imbalance is taken up by the US as an urgent issue and as an example of non-tariff trade barrier between the USA and Japan. Exclusion of foreign contractors is related to the prevalence of the 'Nominated Competitive Tendering System', whereby 10–15 major Japanese firms are selected in accordance with a certain number of pre-qualification criteria, including past construction experience in Japan, by public sector clients to submit a tender. US contractors are now demanding the legal enforcement of an open competitive system in Japan; and the US government (together with South Korea and China) is pushing for the opening-up of the market

on the Kansai New International airport project, as well as on the
bridge and ring road around Tokyo Bay. Possibly as an initial gesture
in response to such external pressures, a $200,000 contract for
reporting on relevant international airport experience was awarded to
the Bechtel Group.

2 Business Strategy

The Structure of Japanese Firms Participating in Overseas Construction

Thirty-three Japanese firms are included in the *ENR*'s Top 250 International Contractors in 1985 as listed in Table 3.2. Japanese firms
which engage in overseas contracting may be divided into the following four groups: (1) general contractors; (2) engineering firms, the
most important of which are in petrochemical engineering; (3)
engineering or construction departments or subsidiaries of machinery
and equipment manufacturers; and (4) trading firms or 'Sogo
Shosha', having no construction/engineering staff in-house. General
contractors are the core of the overseas construction, with more than
50 per cent of total overseas contract awards in the past few years. As
of March 1986, there were 57 such firms registered as members of the
Overseas Construction Association of Japan, Inc. (OCAJI).

The relationship between general contractors and the other types of
firm categorized above should be clarified. General contractors are
heavily dependent upon multinational Sogo Shosha trading firms for
assistance in information gathering, initial project formulation and
planning, major country and/or business risk taking, credit provisions,
negotiating skills, etc.; and they have taken them on as business agents
or major contractors. On the other hand, both engineering firms and
manufacturers frequently invite general contractors to act as their civil
engineering and/or building services subcontractors, or as joint partners where the building and civil engineering portion of a contract is too
large for them to take the risks by themselves. This type of relationship
can seldom be seen at home, where building and civil engineering construction is contracted out separately by clients for sizable projects such
as petrochemical refinery plants, power stations, etc.

The six largest Japanese general contractors are listed and profiled
in Table 3.3. It can be seen that most of them originated as master
builders in the Edo era before 1871, and entered the modern construction business after the Meiji Restoration. All of the firms are publicly-
owned parent companies. Family influences, however, are said to be
observable for all except Taisei. Most of them undertake both
building and civil-engineering work, although the latter became less in

Table 3.2 *The Major International Contractors of Japan: Contracts Awarded During 1985 (millions of 1982 US dollars)*

Rank in ENR	Firm	Type [a]	Foreign	Total	Overseas share (%)
6	Kumagai-Gumi	C	1,947	4,200	46.4
13	Mitsubishi Heavy Industries	M	1,277	1,833	69.6
21	JGC	E	913	1,271	71.8
24	Chiyoda Chemical Eng. & Construction	E	717	1,039	69.0
30	Kajima	C	561	4,434	12.7
34	Shimizu Construction	C	542	4,278	12.7
39	Hazama-Gumi	C	447	2,029	22.0
50	Ohbayashi	C	338	3,831	8.8
53	Kobe Steel	M	315	404	78.0
59	Toyo Engineering	E	282	363	77.8
63	Hitachi Zosen	M	252	n.r.[b]	—
65	Taisei	C	238	4,155	5.7
67	Sasakura Engineering	E	226	226	100.0
78	Nishimatsu Construction	C	201	1,614	12.5
80	IHI: Ishikawajima-Harima Heavy Ind.	M	189	n.r.	—
83	Takenaka Komuten	C	169	3,390	5.0
86	Kawasaki Steel	M	159	456	35.0
94	Ube Industries	M	142	151	94.6
101	Penta-Ocean	C	128	1,261	10.1
111	Hitachi Plant Engrg. & Construction	E	112	821	13.6
115	Sato Kogyo	C	108	1,713	6.3
118	Fudo Construction	C	105	611	17.2
120	Mitsui Construction	C	99	1,691	5.9
129	Mitsui Engrg.	E	89	n.r.	—
134	Tokyu Construction	C	81	1,504	5.4
138	Nippon Kokan KK.	M	79	n.r.	—
139	Maeda	C	51	1,611	4.8
143	Fujita	C	74	1,773	4.2
151	Ando Construction	C	68	n.r.	—
164	Toshiba Engrg. & Construction	E	53	423	12.5
167	Toda Construction	C	49	1,593	3.1
185	Aoki Construction	C	38	n.r.	—
201	Okumura	C	29	982	2.9

Source: *ENR* July, 1986.
Note: [a] Types: C mainly general constructions; M means mainly manufacturing process and power plants; E means mainly engineering.
[b] n.r. means 'no report'.

proportion until the 1986 recovery of investment by the public sector. Takenaka is exceptional in being basically a building contractor, with a comparatively small civil-engineering subsidiary. All of the listed firms are currently attempting to expand their real estate and urban/regional development (equity shareholding) business. Taisei has an industrial housing operation.

The annual turnover of each of these major firms is from three to four billion dollars, which places them among the world's top thirteen (as listed in *Engineering News-Record*) for *total* contracts received at home and abroad. Whereas US and European contractors include in their returns a sizable figure for professional project and construction management (PM/CM), large Japanese general contractors do not engage in project management services. They do, however, have a high percentage of design-build work (60 per cent in the case of Takenaka), which is preferred by private sector clients – an indicator of the contractor's high level of technical capability and trustworthiness in Japan.

All of the firms run full-fledged research institutes and technical development departments that study and research materials, basic and core technologies, and develop production technologies. These include construction robotics, and saleable product lines such as membrane mega-domes, reinforced concrete aseismic high-rise (40 stories or higher) residential towers, etc. The average investment per firm amounts to 6.3–8.8 billion yen except for Kumagai-Gumi's 3.1 billion.

Although the value of overseas contracts received by Japanese general contractors has expanded quite remarkably in the last five years, the ratio of overseas to total contracts of the listed firms for 1985 is comparatively low (5.0 to 12.7 per cent). Again Kumagai-Gumi is an exception, with 46.4 per cent of total contract value overseas. At a general level, a number of incentives and disincentives for Japanese general contractors operating in overseas markets may be identified. These are summarized in Table 3.4, and will be explained briefly below.

Major Incentives for Japanese Contractors Operating Overseas
A number of incentives for Japanese firms to expand overseas have their roots in the domestic economy. First is the fact that total domestic construction investment in Japan has declined appreciably in recent years, with an annual average growth rate of minus 0.4 per cent from 1980 to 1984. The contribution of construction to the GNP has also fallen, from a peak of 24.4 per cent (gross) in 1973 to a trough of 15.5 per cent in 1985. This slump in domestic demand was triggered off by the oil crises of 1973 and 1980. A sudden increase in overseas orders received, after both 1973 and 1980, suggests that overseas work

Table 3.3 Profiles of Major Japanese General Contractors, 1985

	Taisei	Kajima	Shimizu	Ohbayashi	Takenaka	Kumagai
(1) Year of founding	1873	1840	1804	1892	1601	1938
(2) First venture overseas	1959	1955	1958	1961	1961	1961
(3) Capital, 1985 (million yen)	39,231	43,305	35,700	32,616	50,000	27,179
(4) Sales and composition of business, 1985						
Total sales (million yen)	978,607	913,628	1,052,293	797,786	832,126	709,863
Civil engineering, Japan (%)	25.2	34.0	19.3	25.7	2.0	43.4
Building, Japan (%)	72.8	61.4	77.6	70.2	94.4	46.7
Others (real estate) (%)	2.0	4.6	3.1	4.1	3.6	9.9
(5) Annual contracts received 1985						
Total (US $millions)	4,641	4,953	4,779	4,280	3,787	4,692
Overseas (US $millions)	267	628	605	378	189	2,176
Overseas share (%)	5.8	12.7	12.7	8.8	5.0	46.4
(6) Investment in R&D (million yen)	8,500	8,490	8,803	6,384	7,183	3,148
Reg. patents 1973–85	1,164	1,559	1,212	1,035	681	437
(7) Design-build share (%)	ca. 35	n.a.	ca. 35	n.a.	ca. 60	n.a.
(8) Total employees	12,104	13,061	10,090	9,915	9,025	7,916

Source: Annual Reports and *ENR*, 17 July 1986.

was used to compensate for the fall in domestic orders – as was done in previous recessions. During the last two recessions after the crisis, however, the top management of contractors believed, perhaps wrongly, that domestic growth would probably never return as in the past. Thus they have begun to see the overseas market – although apparently very risky – as an important component of corporate strategies.

A second factor relating to the domestic economy is the fact that three decades of rapid growth in domestic investment encouraged leading Japanese contractors to import advanced construction technologies from Europe and North America. These were absorbed and refined, to produce a highly-developed domestic system of engineering, design and construction. At the same time, the demand of Japanese clients for shorter construction times, prompt delivery, high quality of construction and labour safety, together with the absence of construction unions, resulted in leading contractors developing a high degree of expertise in producing good quality work on time. This has gained them an excellent reputation among international development agencies and overseas clients at large, and is a very strong competitive factor in the overseas market.

A third incentive for Japanese contractors to work overseas is the increase in Japanese government's official overseas development aid (ODA). The objective for the period from 1986 to 1992 is to reach a cumulative total of more than $40,000 million – as compared with $21,360 million in the period 1981 to 1985.

A fourth factor that has already been mentioned is the assistance rendered to Japanese general contractors working overseas by the multi-national trading firms, or Sogo Shosha that sustain a major portion of national exports and imports. Closely related is the recent rapid expansion of direct foreign investment by Japanese industry. In 1985 it reached 12.2 billion US dollars, 44.2 per cent of which was in the US, 15.8 per cent in Europe and 12 per cent in Asia. Direct foreign investment encompasses joint development of high technologies with other advanced countries, as well as international production sharing with developing countries. It is expected to expand further as the Japanese yen strengthens and as pressure to reduce the very large external trade surplus increases. Leading contractors are reinforcing their network of US and European subsidiaries so as to be able to assist in direct investment abroad by Japanese manufacturers, service industries and financial institutions.

The increasing internationalization of the Japanese economy has also had an effect upon the development of the construction industry. Overseas business, as well as domestic transactions with foreign firms investing directly in Japan, is the major epicentre of international

Table 3.4 *Overseas Market Incentives and Disincentives*

A Incentive

1 Stabilization of construction investment at home, and resultant decrease in its contribution to GNP

2 A high level of construction technology gained and reputation for quality and prompt delivery from 30 years of continuous growth in the domestic market

3 Ever increasing official development aid to developing countries

4 International and local intelligence networks of Japanese trading firms

5 Internationalization of the Japanese economy, especially the rapid growth of direct investment overseas, and loans by industry

6 Comparatively high economic growth expected in the Asia–Pacific Rim Region, having close ties with Japan

7 1980s economic open-door policies employed worldwide and resultant acceptance of 'with-finance' project proposals

B Disincentives

1 Huge domestic construction market

2 Historical insularity

3 Lack of experience in, and ability to manage, overseas and international transactions

4 An insufficient number and still poor quality of personnel for rapid growth of overseas construction

5 Dominance of Western languages

6 Historical and global spread, and authorized use, of European and American construction contract laws, systems and practices

7 Historically inferior position of Japanese consulting engineers and architects

8 Intense overseas competition among Japanese contractors caused by similar modes of entry

9 Continuing rise in the value of the Japanese yen

10 Intensified competition with European, American and NIC contractors, particularly after the 1983 decline in oil prices.

influences at home that have contributed a lot to the revision of traditional and feudal modes of management. Major contractors are now re-orienting and restructuring industrial and business strategies.

In addition to these domestic incentives, two external factors have contributed to the development of Japanese construction overseas. The first of these is the high rate of economic growth experienced in Asia and the Pacific Rim region – a trend which is expected to continue into the twenty-first century. Japanese economic and political ties with countries within the region (including China, Siberia, ASEAN countries, Korea, Taiwan, Australia, the USA, Canada and Mexico) have become stronger year by year. The second factor lies in the economic and technological 'open-door' policies currently being pursued by China, ASEAN countries, India, Saudi Arabia, Turkey, etc. Even the US, Australia, West European countries and the USSR are very positive about introducing foreign funds for the promotion of high technology to re-vitalize industries, for regional and urban development, and for natural resource development. Large Japanese general contractors, with a high technical and financial capability, have been quick to respond to this opportunity. In the US, Australia and the EEC they are already actively engaged in urban development and/or the renovation of out-moded offices into 'Smart Business Centres', using the most advanced communication and micro-electronic technologies.

Major Disincentives to Japanese Contractors Working Overseas
There are, on the other hand, a number of factors serving as disincentives to Japanese contractors venturing overseas. The volume of domestic construction investment, although stabilizing in recent years, is still enormous; it overtook that of the US some six or seven years ago. The presence of this huge domestic market, has weakened the motivation to operate in foreign markets where risks are substantially greater. Added to this is the natural insularity of the Japanese, which has been particularly apparent in the construction industry. Japanese top managers – most of whom grew up in a period when the volume of overseas business was very small – still generally feel awkward in dealing with overseas-related decisions; at the same time they are hesitant to recruit foreigners to overseas management posts. Consequently, there is a chronic shortage of well-qualified project managers and engineers for overseas construction work. The fact that international construction markets use Western languages, as well as European and American contract laws and practices, creates further difficulties for the Japanese. Education programmes geared to overseas assignment have been underway for many years. These are designed to improve foreign language communication (especially

English), on-site construction management and related claims negotiation, as well as to increase sensitivity to the societies and cultures of the host countries. It is also intended to develop, by secondment to professional offices, a group of specialists who are able to co-ordinate Western lawyers and British or Canadian quantity surveyors, so as to make the most of their services.

A further disincentive to the overseas operation of Japanese contractors is the historically inferior position of Japanese consulting engineers and architects. The number of Japanese designers ranked in *ENR*'s Top International Design Firms for 1985 was twelve out of two hundred. The share of the twelve firms in the total overseas market was only 6 per cent, and 67 per cent of their activities were centred in Asia. By way of comparison, their Western counterparts numbered 168, were responsible for 84 per cent of the market and have been operating in all areas of the globe for many decades. This places Japanese contractors at a distinct disadvantage, both in terms of business information that consultants may supply, and codes and specifications they may use in design.

Finally, the continued strengthening of the yen against other major currencies (the exchange rate of 237 yen to the dollar in September 1985 strengthened to 150 a year later) has served to considerably weaken the international cost competitiveness of Japanese designers, contractors and plant exporters. The former are being forced to rely, against their wishes, upon Japanese Governmental ODA − which accounts for about 10 per cent of Japan's construction work overseas. Contractors, on the other hand, are adopting new types of business strategy (for example, becoming developer-contractors) and moving into new markets (such as the US and Australia). In these attempts they are facing fierce competition, not only from European, American and NIC contractors, but also from other Japanese firms − who all seem to be moving in the same direction at the same time. A further effect of the revaluation of the yen is that Japanese contractors, traditionally heavily dependent upon goods bought in their own home market, are now placing greater reliance upon local or third country procurement of goods and services. The resultant fall in the value of Japanese plant exports was estimated by the Ministry of International Trade and Industry to be in the order of 30 per cent (from $9.6 million in 1985 to $6.7 million in 1986).

Before elaborating on the various strategies employed by Japanese general contractors to cope with the current state of the international market, it may be interesting to outline the way they see their own strengths and weaknesses in relation to both Western contractors and contractors from the Newly Industrialized Countries.

Strengths and Weaknesses of Japanese General Contractors

As already mentioned, Japanese general contractors have an advanced level of construction technology and management and an excellent reputation for producing good quality work on time. They are also heavily involved in technological research, which serves to enhance their engineering, design and construction capabilities.

As regards civil engineering works, they are particularly competent in construction engineering in the fields of tunnelling construction in soft soils, long-spanned bridge construction, large dam construction, large underground-space construction for structures or storage, on-shore/off-shore construction, large prestressed-concrete vessel construction for chemical/biochemical processing storage and nuclear containment, etc. Various types of concrete and prestressed concrete technologies have in fact been developed in-house, with the objective of cost reduction in relation to other kinds of structure. In addition, construction robotics, although still in its infancy, is being enthusiastically pursued in-house; and micro-electron tele-instrumentation is being developed for extreme environmental situations, and for the safety and efficiency of large-scale construction.

As far as building construction is concerned, leading Japanese general contractors have gained a high reputation in the design-build or turnkey approach for quality hotels, high-rise office complexes, theatres and auditoriums, hospitals, educational facilities, housing complexes, etc. All of them have established design offices holding hundreds of architects, structural engineers, building services design engineers, interior and landscape designers and cost planners. Another aspect of their strength is that the designers are supported by technical research institutes and technical development departments in, for example, sophisticated industrialized high-rise construction systems, specifically developed for structural-steel framed hotels, offices and housing complexes; total control systems for air conditioning, surveillance and emergencies, etc. Design-build ratios listed in Table 3.3 attest to their competence.

As regards plant engineering, their competence lies not so much in huge manufacturing plants (such as automobile or petro-chemical plants), but in various types of production plants, such as food, pharmaceuticals and chemicals; high-technology production facilities, such as super-clean rooms for IC tips production; and urban, industrial and radioactive waste treatment, both solid and liquid. Automated physical distribution and storage of raw materials, semi and final products associated with production plants, are regarded as strengths of most of the engineering departments of the leading contractors. New product lines have quite often been developed jointly with manufacturers having specific knowledge and process patents.

In addition to these technical strengths, Japanese contractors also have the ability to arrange financial packages, or to supply finance from their own resources. This puts them at a particular advantage when competing with firms from the NICs.

Despite a considerable amount of overseas experience gained to date, Japanese general contractors still feel that they lag behind their Western counterparts in a number of areas. These may be summarized as follows:

(1) Construction contract management, especially daily on-site management and claims negotiations.
(2) International information gathering, research and project formulation.
(3) Recruitment and utilization of local and third country managers, professionals, and engineers at overseas offices and on sites.
(4) International procurement of materials and equipment, particularly from third countries. (It would appear that the Korean contractors are better off than the Japanese in this respect, as the Koreans are themselves low-cost producers of construction materials and equipment, and such production is often vertically or horizontally integrated with the construction firms.)
(5) Adaptability to foreign societies and cultures. The Japanese are recognized as very shy and passive, whereas Westerners are, Japanese feel, more open-minded and active in receiving foreign ideas.
(6) Familiarity with internationally prevalent Western design codes, standards and specifications.
(7) Professional project management for macro-projects. Since Japanese construction firms still do general contracting based on lump-sum contracts, they have not yet developed project management support systems which enable them to manage projects jointly with their clients.
(8) Total engineering and design capabilities for various kinds of large-scale plants.
(9) The installation of heavy components of process machinery and equipment. In Japan such installation is handled by plant manufacturers themselves, or by their specialty subcontractor, so general contractors generally install building services-related machinery and equipment.

When competing against firms from Korea, Turkey or other NIC countries, Japanese contractors are also very vulnerable (along with their Western counterparts) on the grounds of cost — although this would seem to apply mainly in cases where these countries are able to utilize their own labour-force.

Corporate Survival Strategies of Japanese General Contractors

The 1960–70s were years of exceptionally high economic growth in Japan. Domestic construction investment expanded accordingly and major contractors grew to become financial and technical giants. The first half of the 1980s, however, saw the end of the period of rapid growth and the beginning of a period of much slower growth, characteristic of a mature economy. In addition, an unprecedented rise in the value of the yen took place in the last quarter of 1985. These two factors together have had an unimaginably serious impact on all sectors of the economy and society. In October 1984 the Ministry of Construction formed an industrial research team comprised of academics, economists, officials, corporate planners and marketing specialists related to, and working for the construction industry. In February 1986 they produced a report, 'Vision of the Construction Industry towards the Twenty-first Century'. Stimulated by this endeavour, leading contractors began for the first time to plan survival strategies using modern marketing and corporate planning concepts and methods, including the widely known 'total quality control' (TQC). Most of these concepts and methods were introduced from the US after the Second World War and subsequently refined by Japanese industrialists and academics. The core of the survival strategy of leading contractors, looking 10 to 15 years ahead, is an engineer/constructor plus financier/developer approach. The major components of such an approach are as follows:

(1) Maintaining and Expanding the Traditional Domestic Contracting Base   The first part of the strategy is for contractors to maintain, and if possible expand, their share of the domestic construction market. They should anticipate the potential needs of traditional clients and society at large, while differentiating the service offered through the application of advanced technology, systems and knowhow.

(2) Identification and Cultivation of New Business Opportunities   The second part of the strategy is to diversify into other potentially profitable areas, both related and unrelated to construction. In the former category is the real estate and urban development business. For many years, major contractors have participated in urban-development activities at the request of clients – in exchange for the award of construction contracts. This type of business is categorized as client-led development. Now contractors are initiating projects themselves in order to obtain a stable source of income for the future. They have also begun to invest both financial and management resources in areas unrelated to construction, such as in the leisure/tourism, health and information industries. This development is expected to gain momentum as management acquires experience and employees become more venture-minded.

(3) Reinforcement and Diversification of Business Activities Overseas The third prong of the survival strategy is to expand business activities overseas. In addition to construction activities in developing countries, contractors hope to expand their operations in advanced countries such as North America, Australia and the EEC, where political risks are almost nil, economic activities are far more stable and capital gains on invested properties can be enjoyed. Expansion will be based upon both high-grade engineering and development work. They will not necessarily undertake the actual construction work themselves, but would like to take a prime role in projects as financiers, engineers or construction managers. They also intend to increase their efforts to participate in direct foreign investment programmes of both foreign and Japanese firms − either as consulting engineers, developers, financiers and construction managers or contractors on a turnkey basis − as well as to expand their own direct overseas investment in new areas of business (as outlined under (2) above).

Overseas Strategies of the Six Major Japanese General Contractors
In the light of the general survival strategies outlined in the last section, overseas business strategies of the six major Japanese contractors will now be briefly examined.

(1) Percentage of Overseas Business Taisei intended, several years ago, to increase the overseas share of its contracts to 10 per cent and keep it at that level; in fact it fell to 6 per cent in 1985, most probably due to the chill in the market. Kajima is reported in 1985 to be developing within the next three years a corporate management plan for the 21st century, whereby the overseas share of total sales is to increase from the current 13 per cent to 20–30 per cent. Similarly, in 1986 Shimizu announced a 10-year corporate management programme named SPRING in which the overseas contract composition is targeted at 20 per cent. Takenaka is reported to be trying to increase its overseas business to 15 per cent by 1988. However, in 1985 it was only 5 per cent, and the firm is reported to be positively pursuing overseas projects in an effort to halt the downward trend. Kumagai does not appear to have set any target for overseas work. The 46 per cent share attained in 1985 is not believed to have been planned.

(2) Target Market Geographical areas and the kinds of strategies that have been chosen by Taisei, Kajima, Shimizu, Ohbayashi, and Takenaka are generally alike. They started in Asia, moved to the Middle East, Africa, and back to China and Asia – on the basis of general contracting of heavy civil engineering and building works, design-build projects and plant construction. Now they are shifting many of their resources to Western countries, the US in particular, to pursue real estate development, construction for Japanese firms, civil engineering based construction and consultancy. Uniquely Ohbayashi has been awarded a few shield-tunnelling jobs by competitive bid in the US. Takenaka has established closely meshed footholds covering Western Europe to deal with Japanese firms setting up there.

Kumagai on the other hand is reported to have intentionally avoided head-on collision with the 'big five' listed above, and has established local subsidiaries in Hong Kong and Indonesia in 1972, Taiwan in 1974, Australia in 1982 and Canada in 1984. It has generally been running its overseas business in a unique manner. For example, its business in Australia expanded suddenly in 1982 as a result of making loans to local developers in return for construction jobs.

Vast geographical areas that remain comparatively or completely new to Japanese general contractors are Australia, India, Central and South America in the former category, and the USSR, Eastern Europe and the Arctic Region of North America in the latter category. In view of current diversification strategies such natural resource-rich countries as Australia and South America must be very attractive: Kumagai has already participated, in 1986, in the 'Very Fast Train Project' connecting Sydney and Melbourne. Aoki, a medium-sized contractor, has invested and diversified in Brazil. When the political relationship between Japan and the USSR is revised, Siberia will be a tremendously challenging market both in terms of business and cold-region technology.

(3) Revision of Responsibility Systems for Overseas Business One of the characteristics of Japanese firms in general is that the majority of decisions are made on a group-consensus basis. This system works reasonably well in Japan. But the development of initiative, clear-cut leadership and quick responsiveness, all of which are regarded as merits of modern managers by Western standards, is inevitably hindered by it. Hence, overseas business transactions are impaired. Now is the time, however, for decisive action overseas as well as at home in order to outdo international competitors or to formulate joint ventures with foreign firms. Both competition and joint venture formulation demand initiative, leadership, quick responsiveness and responsibility of managers in charge. Japanese contractors have therefore begun seriously considering how to overcome the shortcomings of their decision-making systems.

One of the changes some of the contractors have made is to delegate the authority for signing contracts and responsibility for raising profits to a lower management level. Kajima assigned three regional general managers at board level, and Shimizu eight regional managers at middle-management level, within their respective international divisions in 1985. Kumagai for many years has had a different approach, that of requiring overseas branches to report directly to the company president, along with the central international division and its overseas offices, and subsequently increasing the number of branches. It is anticipated that such actions will speed the development and increase the number of younger managers – who should be the most valuable asset of any firm.

(4) International Finance Proposal A couple of examples of a Japanese contractor taking full advantage of the possibilities opened up by the international provision of finance may be cited. In one instance, it is reported that the firm joined forces with an influential local developer and agreed to a loan. When the developer obtained a sizable urban redevelopment project, the Japanese contractor undertook responsibility for the construction work and then subcontracted it to local contractors. In another instance it is reported that it invested directly in a development project, that it became the general contractor of the construction, and that, at an appropriate point of time after the completion of the project, it will sell out its portion of the property to gain a profit.

Project Staffing
Because Japanese employees are generally employed on a lifetime basis, the core members of the overseas construction team, such as project managers or construction managers, office managers, construction engineers and administrative officers, are mostly drawn from those well-experienced at home. Office staff such as clerks and secretaries are recruited locally. In recent years joint ventures with local contractors have been increasing in line with the indigenization policies of host countries. In such joint venture projects, engineers and office staff assigned by the local contractors quite often work under Japanese counterparts, mostly because of the technological and managemental abilities required for large and technically complex projects.

The number of Japanese employees with overseas experience, known as 'overseas veterans', has been growing. However, since overseas activities are also expanding and employees tend to dislike leaving families behind for any length of time (despite the provision of substantial overseas allowances and paid leave), the availability of these veterans willing to go out is still tight. Company-wide rotation and re-rotation programmes for overseas assignment are incorporated into career development plans. In many cases project teams are formed by mixing overseas veterans and home veterans of the specific types of construction required for particular projects, say, tunnelling, high-rise construction, etc. Naturally those who are new to, or still inexperienced in overseas assignments are more likely to face linguistic and cultural problems, as well as difficulties in dealing with the client's representatives, consulting engineers, subcontractors and locally recruited staff. Japanese contractors have recently undertaken a number of measures designed to alleviate overseas staffing problems. First, when taking on new recruits, language abilities are carefully assessed. Follow-up training in English is offered to nominated employees; this may continue over a period of many years. Chinese, French and Spanish however are also currently being proposed.

General overseas education programmes also exist for the study of business administration, Western law, construction contract management, project management and architectural design. All of these are at the graduate level, combined with several months of foreign language brush-up training, and conducted at major colleges and professional offices abroad. Such programmes may be open to all employees or to specially selected nominees. Generally five to ten a year are selected from each company.

Most companies also run their own education programmes for on-site operations overseas. Groups of employees are educated intensively at the companies' education centres for a month or more. The contents of the programmes are oriented towards site operation, are extremely practical, and include language and cultural studies. At an industry level, basic overseas construction education programmes are provided by the Overseas Construction Association of Japan, Inc., the Engineering Advancement Association of Japan (ENAA) etc. and are open to all member companies.

In addition to the training of Japanese employees for overseas assignment, the recruitment of local or third country professionals by Japanese general contractors working in developing countries seems to be increasing – a movement which has no doubt been stimulated by the strong rise of the yen. According to the Ministry of Construction, the employment ratio of foreigners to Japanese on the business staff of Japanese overseas subsidiaries may be as high as 85 per cent in the Middle East and ASEAN countries.

As regard the recruitment of labour, it should be remembered that Japanese general contractors do not have foremen within their own organizations. In domestic building work, almost all work except for a limited portion of general items is subcontracted. In civil engineering, subcontractors generally work continuously for specific general contractors. The subcontractors' engineers thus play the role of foremen for the general contractors. When starting work in a new area, contractors are known to evaluate carefully the relative advantage of hiring labour directly, or subcontracting. In the Middle East, for example, where there is scarcely any local labour or subcontractors, one Japanese contractor began in the 1970s by asking Japanese subcontractors to supply foremen to administer and control labour hired directly from third countries. As the cost of the Japanese foremen rose, the general contractor switched to foremen supplied by the subcontractors of the third countries themselves. They then began to train foremen from the Chinese labour corps contracted on a direct hire basis until they eventually gained sufficient experience for an increasing portion of the work to be let out on a subcontracting basis.

Technology Transfer in Construction

An increasing number of local subsidiaries have been established in recent years by Japanese contractors overseas, as indigenization policies of the host countries become more strict and the locally contracted portion rises. Joint ventures between Japanese and local firms are also increasing. Thus more and more requests are made on a government to government basis or on a partnership basis for the education of construction engineers, design engineers, office administrators and the like.

In the area of construction engineering and design, Japanese contractors are well advanced. They provide training in these areas for a number of students, mostly from other Asian countries, either directly or at the request of Japanese governmental agencies in charge of technology transfer. Japanese general contractors, with the assistance of their subcontractors also provide training at a project level to local foremen or craftsmen. This may be carried out either locally, or in Japan. In the field of overseas project management, however, the Japanese are somewhat behind their Western counterparts. It is therefore doubtful if they can contribute much to the education in this field of engineers and managers from countries with strong Western influences. Japanese contractors who have worked in China, however, are said to feel that the Chinese may be able to learn from them when they come to modernize their construction management systems.

3   Japanese Government Policy Towards Overseas Contractors

Japanese construction firms have a number of natural advantages when operating in overseas construction markets. However, they also face a number of problems. These stem largely from the fact that they have a relatively short history of commercial operation overseas and have tended to concentrate their operations upon developing regions with high political risks. Their major problems are the following: (1) the continuing shortage of management and engineering staff for overseas work; (2) the weakness of consulting engineers and architects, a large number of whom have little overseas experience and local know-how and lack all around capacity; (3) the absence of financial support to secure orders and a guarantee system for losses due to political upheavals; and (4) the need for government initiatives to increase international competitiveness. To cope with these basic problems, various construction promotion/protection policies have been adopted by the Ministry of Construction, the Ministry of International Trade and Industry (export insurance) and the Ministry of Finance (tax and loans). These are applied in the same way to consulting design firms and machinery/materials suppliers as to construction companies.

Overseas Construction Promotion Policy by the Ministry of Construction
Specific measures undertaken by the Ministry of Construction are as follows:

(1)   The establishment of an overseas construction promotion fund to supply low interest credit for pre-bid feasibility studies.
(2)   A financial guarantee system for overseas construction projects by overseas construction companies and consulting firms.
(3)   An 'Infrastructural Facilities Investigation' and 'feasibility studies for construction projects' to find out and to form suitable projects in developing countries.
(4)   An overseas construction technology development project to develop the appropriate technology of construction in conformity with natural and socio-economic conditions of developing regions.
(5)   A training system of consultants and project managers in charge of overseas construction projects.

Overseas Construction Promotion Policy by the Government as a Whole
In addition, various policies for the promotion of overseas construction have been set up by the government as a whole. These are described briefly below:

(1) Institutional finance: yen credit is provided by the Overseas Economic Co-operation Fund of Japan, and export and technology suppliers credit by the Export–Import Bank of Japan. Such credit is not available, however, to individual corporations, but only for collective or joint-venture projects (see below).

(2) Export insurance system. The Ministry of International Trade and Industry (MITI) has an export insurance system incorporating the following: general export insurance, export proceeds insurance, exchange risks insurance, export bond insurance, overseas investment insurance and technical services supply insurance (TSSI). The last (re-named in 1983 from 'Overseas Construction Works Insurance' set up in 1979) is the most important from the construction industry's point of view. Its objective is to insure equipment, etc., to be used for the supply of technical services against the impossibility of collecting invested costs due to political or commercial risks. It may be applied when a project has been interrupted for more than six months by war, etc., such as the Iran–Japan Petrochemical Construction Project now under examination. In practice, the value of TSSI is limited, due to excessively rigid requirements of contract and risk-recognition, such as the refusal to recognize the Iraq–Iran conflict as 'war' (see below).

(3) Tax reduction system. Preferential treatment is afforded to Japan's export industries not by tax credits but by the reduction of taxable income. For overseas consulting firms 20 per cent of total income is deductible prior to taxation; this is not only to counter high political risks but also to enable them to raise research and development funds and to foster business vitality. Profits of overseas constructors are regarded as contributions to 'Development Assistance'; and a proportion may be included in reserve funds for future losses and be exempted from taxation for five years.

(4) Official development aid (ODA) is administered by the Ministry of Foreign Affairs and may be on a bilateral basis, in the form of a grant or direct loan; or a multilateral basis, in the form of financial support for international organizations.

(5) Technical co-operation by the Japan International Co-operation Agency (JICA). JICA has a number of functions including: invitation of people from developing countries for technical training in Japan, the dispatch of Japanese experts and volunteers abroad, the dispatch of survey teams to help in formulating development plans and projects, the provision of grants for equipment, and the extension of project-type technical co-operation.

(6)   The government provides support for education of Japanese children living overseas, as well as for travelling medical services in developing countries.

(7)   'Construction Attachés' have been dispatched to 23 countries, with the responsibility for economic and technical co-operation and for the gathering of information related to construction projects.

(8)   Japan–China construction technology interchange project. A joint-venture was proposed by the Chinese government in October 1978. Under this technology transfer agreement Japan Steel Corporation, Japan's foremost steel company, undertook the construction of Baoshan Steelworks in Shanghai, China.

(9)   Direct governmental intervention. In view of its policy of 'free trade' the Japanese government does not exert pressure on foreign governments or international agencies to award contracts to Japanese firms. It does sometimes make use of ODA or other means to provide financial support to a foreign government, in order that it may participate in a national project (e.g., support for the Bosporus Bridge). Apart from ODA, there is one example of an international tender that required financial assistance – that of the Hong Kong Subway Works (first phase in 1973). OCAJI appealed urgently to the government, and finally the Export–Import Bank of Japan decided to provide a deferred payment loan to Hong Kong, which ensured the success of the Japanese joint-venture's tender. Since then, however, the Ex–Im Bank has supplied no credits for service exports, although they are still available for plant exports.

Government Support and Future Prospects

On the basis of government reports on the future prospects of Japan's overseas construction industry, as well as opinions expressed in interviews, a number of observations would seem to be in order.

First, Japan's institutional financing system has some positive aspects; but unlike its counterpart in Britain or France, it does not supply credit to individual corporations. This works to the disadvantage of Japanese contractors who are not involved in joint-ventures for overseas contracts.

Secondly, as regards Japan's export insurance system, the 'Technical Services Supply Insurance' seems to give little benefit to Japanese construction firms; in fact until quite recently there was no recorded incidence of an 'insurable risk' for which government actually paid benefits to the insured firms. For example, the fighting between Iran and Iraq has not been recognized as 'war', but simply as a 'dispute'. Consequently, ordinary overseas construction firms have no reason to affiliate to the export insurance scheme (unless required to do so by a supplier's loan from the Export–Import Bank). From now on, it is hoped that Japan's export insurance systems will be internationalized and integrated into MIGA (Multinational Investment Guarantee Agreement) set up to protect companies against political instability. In such cases an Assessment Committee, whose job should be to recognize whether an accident is to be covered or not, for war, guerrilla activity or strikes, etc., is indispensable. A long-term export insurance such as the French have, one that reportedly gives guarantees for ten years, is highly desirable. So far as the training of engineering and management staff is concerned, there are no competent planners of overseas public works projects in the major private construction corporations of Japan. In government organizations, on the other hand, an internationally experienced planning staff has been accumulated. In this age of turnkey projects, training and interchange of such staff between government and private industry is urgently required. Flexible measures are also required to enable contractors to cope with the problems of 'counter-trade' (or payment by oil, or other resources).

Finally, it may be concluded that Japan's construction industry is now facing the biggest transition since the end of the Second World War. The traditional practices of contracting a project, building it, and receiving payment for the work are no longer sufficient for construction companies to survive either at home or abroad. The following three constraints are likely to be of vital significance in the development of Japan's overseas construction industry in the coming 'Pacific Age': (1) the impact of the rapidly strengthening yen since September 1985, which is dealing a heavy blow to Japan's overseas competitiveness; (2) declining oil prices in the oil-rich developing countries, which is reducing Japan's overseas markets; and (3) defective financial insurance policies in Japan – which will have to be changed.

4

Italy

ALDO NORSA

1 Introduction

Italian contractors abroad had their highest share of the world market (10.6 per cent) in thirty years in 1985, with their 1986 share only slightly smaller (10.0 per cent). Confronted with shrinking international demand (a fall of 49 per cent since the peak achieved in 1981 according to *Engineering News-Record*), the top Italian contractors have also had their second best year (after 1981) in terms of value of new contracts ($8,714 million) acquired in as many as 92 countries. This chapter tries to explain how Italian contractors have managed to obtain a performance which makes them third in the world for the third consecutive year (behind the USA and Japan) and how they prepare to keep their share of a market which appears to be shrinking further.

Italian official figures are fragmentary since the only consistent source is the civil contractors' association (ANCE) excluding all non-civil firms (those building industrial and power generating plants). The source which is used here is the *ENR* listing of the top 250 international contractors, which offers comparable (although not totally homogeneous) data for six years: 1980–85.

Recent Trends

According to *ENR*, Italy had 28 top international contractors in 1985 (their average number had been 21 in the five preceding years). Expressed in 1982 constant dollars, the $7.8 billion awards of 1985 show a 23 per cent increase over the previous year (26 firms), a 20 per cent decrease on the record year 1981 (18 firms), and a slightly better performance ($+2$ per cent) than the average of the preceding 5 years (1980–84) (see Table 4.1). This performance can be fully appreciated

Table 4.1 *Italian International Construction, 1980–86*

Year	Number of Italian firms among the world's largest 250	Contract awards in 1982 dollars (millions)	Percentage share of contract value awarded to the 250
1980	24	7,191	5.7
1981	18	9,818	7.1
1982	15	7,847	6.4
1983	20	6,962	7.7
1984	26	6,360	8.5
1985	28	7,802	10.7
1986	35	6,463	10.0

Source: *ENR* annual surveys of the world's top 250 contractors.

when compared with a 34 per cent decrease between 1985 and the 1980–84 average for the top international contractors as a whole (with only Japan doing better than Italy). Moving from sixth in the world at the beginning of this decade to third in 1984 and 1985, the Italian share of the market has increased from a 7.5 per cent average in 1982–4 (7.1 per cent in 1980–84) to 10 per cent.

As far as civil engineering contractors were concerned (the only group which can be analysed in detail) their role appeared uneven in the six-year period: they accounted for over half the value of foreign awards in 1980, descending to a minimum of one-quarter of Italian awards in 1982 and 1984, amounting to approximately one-third in 1981, 1983 and 1985. Expressed in 1982 dollars, civil contracts had their record year ($3.7 billion) in 1980 and were down by 27 per cent (to $2.7 billion) in 1985. Their performance was just a little weaker than non-civil contracts, which had a stable value of awards in the preceding five years (1980–84).

A comparison between *ENR* data and the official figures from ANCE concerning civil contractors has limitations since the former refer to just 15 contractors and the latter to 36. Furthermore Italian firms report differently to the two organizations: contracts awarded to all their subsidiaries (*ENR*) versus the awards just to the parent company (ANCE). This notwithstanding, the Italian association announced for 1985 the best performance abroad since the record year 1981: $2,671 million ($2,391 million at 1982 values). This result in constant dollars is nearly 45 per cent lower than in 1981, although more than double that of 1984. Like the *ENR* data the amount matches the previous five-year average ($2,389 million).

Table 4.2 Turnover and Contract Awards of Nine Leading Italian International Contractors, 1983 and 1985
($ millions, 1982 prices)

	Foreign contracts			Total contracts		
	1985 (Turnover)	1983 (Awards)	1985 (Awards)	1985 (Turnover)	1983 (Awards)	1985 (Awards)
1 Sadelmi	477.3	1,927.8	1,575.1	502.4	1,945.1	1,610.6
2 Impresit/Impregilo	416.2	1,590.8	1,757.1	514.2	1,737.2	2,013.6
3 Snamprogetti[a]	353.4	914.3	662.5	490.9	1,491.8	841.5
4 Saipem[a]	569.8	719.0	739.7	712.2	803.7	814.6
5 Cogefar	171.4	251.5	86.2	323.4	438.4	323.2
6 Astaldi	202.1	222.8	136.2	288.7	222.8	209.3
7 Ansaldo[a]	272.6	167.5	147.7	1,362.8	299.5	448.5
8 SICEL	122.2	124.6	120.9	147.3	134.9	137.5
9 Italimpianti[a]	526.1	52.9	1,047.9	618.9	125.3	1,243.5
10 TOTAL	3,111.1	5,971.2	6,273.3	4,960.8	7,193.7	7,642.3

Source: Engineering News-Record, July 19, 1984; June 12, 1986; July 17, 1986.
Notes: Impresit and Impregilo have been combined since Impregilo consists of Fiatimpresit together with Girola and Lodigiani of Milan.
Cogefar, Astaldi, and SICEL are civil contractors.
[a] State operation.

Table 4.2 shows that for a given sample of ten large firms, the real growth in contract awards from 1983 to 1985 was only 5.1 per cent, while the entire set of Italian firms grew by 12.1 per cent. As we shall see later, recent Italian growth was due to awards to smaller firms and new consortia, not due to the consistent growth of the giants. Indeed, if Italimpianti had not won an enormous contract for a steel pipe mill in the Soviet Union, awards for these ten would actually have declined.

Also evident in Table 4.2 is the usual large difference between actual turnover and contracts awarded, as reported in *ENR*. Italians do not double-count reported figures with two firms reporting the same project as either general contractor or construction manager as Americans do, because construction management is not the custom. But Italians have a penchant for forming consortia that in turn combine to form others in complex ways that will be described later. Double counting occurs when each level reports the same award as accruing to it. For example, the combined awards reported by Impresit and Impregilo in 1983 came to $1.6 billion (1982 dollars), but the amount of work that was duly reported two years later as turnover was only $416 million. The two-year lag is not that precise, and again the case of Italimpianti weakens the disparity. Nevertheless, for the set of ten firms the $3.1 billion of work carried out in 1985 was barely half as much as the $6.0 billion in contracts reported as won two years earlier. In a recovering market, one would not have expected that without double counting in awards compared with turnover. The *ENR* data, we may say, overstate the volume but adequately reflect changes.

As for longer term trends in civil contracting, Italy has experienced, as have her main competitors, a 'boom decade' between 1973 (the first oil crisis) and 1981, with awards rising more than four-fold in current dollars (from less than $1 to more than $4 billion). This growth appears irregular: after a period of even increments between 1973 and 1977 (to some $3.5 billion) there were ups and downs until 1981, followed by a severe recession in 1982–4. According to the first ANCE unofficial estimates, 1985 is an exception (equal, however, in current dollars to the values of 1978 but lower than 1977, 1979, 1980 and 1981).

Beginnings
The early history of Italian construction abroad can be traced back to the 1920s when the prospects of works in Africa (in connection with the country's colonial adventures) were such that the Fiat group decided to found Impresit (1929) with the explicit task of providing management and financial support for overseas contractors.

A real blossoming of Italian contracting on a world scale happened in the mid-1950s, after the country had completed reconstruction and had accumulated unique know-how in the erection of dams and hydroelectric power stations all over the Alps. In 1955 the International Contractors Section was established as part of ANCE and 1956 marked a major breakthrough: the prestigious Kariba dam, in what was then British Rhodesia, was awarded to four leading Milanese contractors: Impresit, Girola, Lodigiani and Torno. This was the beginning of a series of achievements, especially in the construction of hydroelectric schemes but also in other major civil works, and marked the creation of Impregilo, a permanent worldwide joint venture between the first three firms.

Activity abroad remained mostly confined to the civil sector (with the notable exception of work in power generation) until the first oil crisis of 1973. Italy had by then developed as a major industrial power and started exporting know-how in an ever-widening range of technology for plants (with turnkey contracts more often than not). By the early 1980s factory building along with related infrastructure (power generation, electrical and telecommunications) had become the dominant contracting activity.

A report called 'Italian Engineering and Construction 1986', based on a survey specially carried out by *ENR*, described the 129 most relevant contracts acquired by 64 contractors in 76 countries. The number of countries where major Italian firms had found new work in 1985 was even higher: 92 in all continents (according to an *ENR* top international contractors survey): 17 in America, 14 in Europe, 13 in the Middle East, 11 in Australasia and 37 in Africa.

In mid-1986 no fewer than 100 Italian contractors of repute were doing work overseas; the directory prepared for the *ENR* special report of 1986 lists 80 of them: 40 civil engineering contractors, 28 non-civil contractors, 12 construction engineering firms (offering design services besides execution), to which should be added the engineering divisions of a number of manufacturers not yet established as firms in their own right. Of the 80 contractors, 30 declared a 1985 turnover of $100 million or more and 25 of $50 million or less.

As far as civil construction was concerned, it is estimated that out of some 100 general contractors (not all big firms) who have worked abroad, only about 50 were active at that time. According to ANCE, in 1985 they operated at 257 construction sites (99 of which awarded in that year) in 56 countries (in 3 for the first time and in 32 with new work). The degree of concentration among overseas construction firms is remarkable: the top 4 won 61 per cent of the contracts in 1985 (against a 60 per cent average in 1980–82); the top 8 won 79 per cent (76 per cent previous average); but it should be remembered that the biggest, Impregilo, comprises three firms.

To fully appreciate the importance of overseas construction for the Italian economy the key figure is 15 per cent: 15 per cent is both the proportion of foreign work in total construction activity and also its value as a share of total exports of goods and services. Furthermore, 15 per cent of insurance accorded to foreign activity is for construction.

More difficult to assess than the actual value of new contracts, normally inflated during construction, is the income earned for the country from construction abroad. Nomisma, an economic research centre in Bologna, estimates the amount of 'repatriable income' at between 2–3 per cent and 12–15 per cent of final contract value (depending on the type of work performed: turnkey having the higher ratio). The so-called 'induced' portion (as an incentive to the export of other Italian goods and services) cannot be quantified: it has however been shown to be positive by Nomisma (with a two-year lapse) through a regression exercise on 12 developing countries where important projects were under way in 1982.

The Nomisma research estimates that around 18,000 Italians were employed abroad in 171 surveyed construction sites in 1982 (a figure for non-civil projects is not available): this represents barely 1 per cent of all registered workers in the national construction industry. As practically all Italian expatriates are skilled (and many of them are technicians and professionals) their salaries are estimated to account for as much as 5 per cent of the national construction payroll. As the number of expatriate workers has kept diminishing, an educated guess reduces the 1986 figure to less than 15,000 (a parallel reduction of manpower in construction in Italy has kept the ratio even).

The Italian ability to shift from one area to the other is probably the single best explanation for the country's increased share of world's contracts in the first half of the 1980s. Emphasis on industrial plant construction has developed parallel to the time-honoured skills in civil works, as more and more manufacturers (some of them even small, peripheral, but highly specialized and already successful exporters of products) have discovered this opportunity, and have turned themselves into constructors, either through establishing an *ad hoc* division or through the initial help of engineering firms.

Integrated development projects (with more and more engineering services involved and infrastructure of a level such that they can be executed locally) are also being sought, as the best way to exploit the opportunities of fast-growing Italian aid financing. The recent comprehensive rural development programme for Senegal allows, for the first time, civil and engineering firms (typically separated in the domestic market) to join forces in a consortium, to share work and to acquire know-how.

Support from the Italian government to overseas contracting has been on the increase since 1977 (when coherent legislation reforming export procedures was introduced and gradually implemented) and 1981 (when foreign-aid programmes finally became substantial and *ad hoc* credits were geared to the needs of commercial penetration). Although Italian administration is known for inefficiency, the drive of the industry seems to be such that even informal promises of support are sufficient for preparation of winning bids. The good reputation that the Italian contracting industry enjoys nationally is the best assurance that 'l'intendance suivra' once the contract is obtained.

The Italian market, despite EEC regulations, is not in practice open to foreign competition, due to procedural barriers and logistical obstacles. An interesting recent development is that the largest overseas European contractor, the German Philipp Holzmann, has opened an Italian branch and in 1985 won an initial $15 million motorway contract. In the immediate future it is expected that major European construction firms will co-operate to boost investments in infrastructure on a continental scale, as a counterbalance to the weakness of Third-World markets. (The tunnel under the English Channel and the Messina Strait bridge are the two most prominent examples of projects needing international private financing, hence involving contractors from all over Europe).

2 Business Strategy

It should be stressed that there are two facets to Italian contracting abroad: civil and non-civil firms (which are responsible for one-third and two-thirds of 1985 contracts respectively). Very little co-operation exists between these types of firms, even when they belong to the same groups: the former are represented by a single association, ANCE, which performs as a rather effective lobby through the International Contractor Section; the latter regroup in various organizations: OICE (which also includes design firms), ANIMP, and UAMI. From the point of view of ownership there is yet another set of distinctions among private, public and co-operative firms; the latter have their own association, ANCPL, but are only active in the civil sector at present, with no more than five major names.

A few corporations (industrial and/or financial) are active in overseas contracting through specialized firms: the two large state-owned holdings – ENI and IRI – and some big private groups: Fiat, Montedison, Ferruzzi, Acqua Marcia/Bastogi.

ENI, the national oil corporation, owns three contractors operating in related fields (although actively diversifying): Saipem, Nuovo

Pignone, and Snamprogetti. IRI, the largest European non-oil corporation, has a civil holding – Italstat – controlling two overseas contractors – Italstrade and Condotte – and owns the major plant-maker Italimpianti, the specialist builder of power stations, Ansaldo, and Sirti for telecommunications. Fiat, through its newly created holding Fiatimpresit, owns the engineering/constructor Fiatengineering and two major overseas civil contractors – Impresit and Imprefeal – besides participating in Impregilo and in Imprestirling (with Federici). It also controls directly SNIA/BPD for chemical process plant engineering. The chemical corporation Montedison operates through the constructor Tecnimont (and Montedil as civil contractor). The commodities multinational Ferruzzi controls three smaller civil overseas contractors: Cisa (active in manufacturing factories and agricultural buildings), Gambogi and Almagia (with CMC, specializing in marine works).

Besides these industrial groups, limiting our analysis to the ten major civil and non-civil contractors overseas, there is only one financial group (as banks are not allowed to control such firms) which is active in construction: Acqua Marcia/Bastogi which owns the construction giant Cogefar (and the engineering firm CTIP) and also has a family alliance with the civil contractor Romagnoli. As far as co-operative firms are concerned, their close political ties in the domestic market appear loose overseas where no significant success has yet been obtained outside the traditional civil-works market. Acting independently and allying with private and public contractors is the largest co-operative builder in Europe: CMC.

Apart from major groups, interesting synergies are obtained through a solution peculiar to Italy: permanent 'joint ventures', but acting as full-fledged contractors, between firms with compatible know-how, working exclusively abroad. The two best-known examples are: GIE (Ansaldo, Franco Tosi, Riva Calzoni) created in 1953, for power stations, and Impregilo (Fiatimpresit, Girola, Lodigiani) born in 1960, specializing in dams, harbours and large civil-engineering works.

All the other firms in the 'top ten' list are privately owned, generally parent companies of larger integrated groups, all fully Italian, with only the exception of Technipetrol, established in 1969 at the time of the oil boom and controlled by the French engineering giant Technip.

Some major civil contractors took on international work as early as in the 1930s (Impresit being the leader then). Others started in the 1950s (Astaldi, Cogefar, Imprefeal, Recchi); and a minority began in the last 15 years (CMC, Sicel). The history of non-civil contractors is shorter: their presence abroad dates from no longer than 15 years ago (when Third-World countries became good clients of turnkey plants and installations). Notable exceptions were firms active in the electrical and power fields, such as Ansaldo (and its offspring Gie), Sae and Sade/Sadelmi.

As far as relations with consulting and design firms are concerned, an Italian peculiarity is the weakness of such independent firms in international markets. This weakness is a result of their lack of opportunity to grow in the national milieu. An archaic law limits designing services to private, unincorporated professionals (so that in fascist times Jewish architects and engineers would be conspicuous and easy targets for discrimination). Some of the major contractors (typically non-civil) have a design unit, as a service which they can also provide separately, and are thus called 'construction/engineering firms'; these include Ansaldo, CTIP, Fiatengineering, Snamprogetti, Technipetrol, Tecnimont. Others are just accustomed to working with consultants from any country, a flexibility which is a reason for their versatility and worldwide success.

Types of Contract

The types of contracts which Italian firms seek vary but generally include, both in the civil and non-civil sector, full responsibility for the erection of the facility (there are therefore no sub-contracts, except for specialist contractors, and no construction management). The philosophy is that know-how in construction is precious and can easily be lost by simply offering services without being directly involved in site work. This is true both of traditional civil contractors – who are jealous of their building talents and, if the project is big, prefer to join forces with firms of their size – and of non-civil contractors, who rely on a number of manufacturing firms behind them (with more or less formalized ties) supplying parts essential to the quality of the facility. In this case turnkey contracts are generally preferred as the competitive edge has to do with proprietary rights. (Such is the case of Italimpianti, which has been awarded all seven contracts for pipe mills in the world in 1985–6.)

A survey shows that Italian builders are individualists, 88 per cent preferring to work alone. Nevertheless, all types of 'joint ventures' are practised by Italian firms: among themselves – temporary or permanent – with 'host country' firms and with contractors from a third industrialized country. The goal, however, is to retain sponsorship of the project. These joint ventures are an especially common practice in large civil engineering projects: a typical example is the most famous contract of 1985, the bridge on the Bosporus, with Impregilo (itself a joint venture) sponsor of a consortium with Japanese and Turkish companies.

Relationships with host countries are crucial and major firms try to be permanently present once the first contract has been obtained. This stability is especially demonstrated in civil works: out of 56 countries where the Italians worked in 1985, further contracts were obtained in 32 and only 3 were 'hosts' for the first time. In any event, due to severe currency regulations, it has been customary to establish subsidiaries in all countries (even in those where local laws do not require it). As a result there are more than a hundred countries where Italian contractors are located with such stability that even serious wars do not cause them to leave (an example being the Bandar Abbas port in Iran).

In terms of overseas bidding and contracting policies, a stable presence in so many countries is a key to the success of Italian contractors: it allows them earlier information before bids are officially requested. At times of mounting international competition (and especially since the beginning of the 1980s) the Italian industry has been supported by a stronger commitment from its government; not only with bilateral aid and co-financing schemes but also through numerous counter trade deals with third countries.

Finance and Counter Trade

The involvement of Italian contractors in the provision of finance for projects appears to be the highest in the world. According to *ENR*, in 1985 (but this was also true of 1984), contractor-financed work amounted to more than $1.9 billion (i.e., 22 per cent of all awards); this was followed by the Japanese ($1.75 billion, i.e., 15 per cent of their awards). Project financing is thus common practice, aided by the fact that not only national but also foreign banks are allowed to benefit from state support when they grant loans relating to Italian export contracts. A successful credit package was assembled for the Bosporus bridge bid − jointly with the Japanese − by a consortium of Italian banks led by the American Chase Manhattan. A flexible use of currencies in financing schemes is another key to success (as proved by the biggest contract of 1985 − the Voljski steel works in the USSR − initially denominated in dollars and then transformed into European *ecus* by a consortium led by Banca Commerciale).

As far as counter trade is concerned, although it is recognized merely as an expedient, cyclical solution, it is being used more and more because of the serious financial difficulties of many Third-World countries. Besides the two state-run corporations − ENI and IRI − which almost have their own foreign and commercial policies with selected nations and thus organize counter trade systematically, other firms are helped in accepting this form of payment, either in the context of government-level bilateral agreements (such has been the case with the USSR) or when problems of payments arise for projects which have already been awarded (this is presently the case in trading with Iraq).

Another example of innovative financing is provided by Italy's agreement with the World Bank: in 1982 Italy was the first country to sign a protocol for co-financing projects in developing countries worth $450 million for a three-year period, a scheme renewed in 1985 for three more years and the same amount. Although there is no direct connection, World Bank sources note that Italy's solidarity with the international institution is justified: for every lira contributed to the Bank's programmes Italian contractors seem to obtain 4.12 liras worth of awards.

Fiatimpresit and Ferruzzi

Apart from the general business strategy of the Italian industry, the programmes of a few groups are worth examining in their own right because of the way that they respond swiftly to changes in the world market. The creation in late 1985 of Fiatimpresit, the holding company of the Fiat group's civil engineering sector, demonstrates both how important project financing has become and how crucial multidisciplinary complementarities are for integrated project development. Capitalizing on 55 years of experience (Impresit was founded in 1929 and has been involved in major works overseas since 1933), this new holding company has the role of planning, co-ordination and control since its managers believe that the integration of roles is a factor of success. The philosophy is that, when dealing with Third-World countries, an industrial group should be able to provide complete development projects, from conception through realization to operation. This role, referred to in some literature as 'project champion' (a more entrepreneurial extension of 'project manager'), means that Fiatimpresit adds to the range of services offered: they integrate the activities of promotion and development, project financing, technical and economic feasibility studies, preliminary studies and detailed project design, management and execution of works, starting-up, training of personnel and eventually even running the works. Good examples are projects of land reclamation from the desert in Bakolori (Nigeria) and in Mujib (Jordan).

Another crucial aspect to Fiatimpresit's business strategy is the realization that the Third World (where the firm has two-thirds of its turnover) is too risky, while the Italian market is too small for its operations. Hence a plan to develop a European 'domestic market' by promoting big infrastructural projects to unify the continent physically. Fiatimpresit is actively proposing institutional and financial solutions to guarantee, through modest sums of public capital at

national and EEC level, a large mobilization of private funds
to finance projects in the interest of Europe as a whole.

Another group starting to work along similar lines is Ferruzzi (by
now a multinational power in such basic commodities as wheat and
sugar). Logistics are seen as the crucial problem for development, thus
Ferruzzi's competence in trading commodities can be integrated with
civil and industrial works. The Italian aid programme (and especially
its food division) has provided the best opportunity to develop this
know-how: Ferruzzi has made an agreement with the government to
provide commodities at short notice anywhere in Africa where
problems arise, and to develop those facilities (ports, roads, silos, etc.)
that are found lacking. Hence it has pursued an aggressive policy of
acquiring specialized contractors – from its original stronghold Cisa,
Almagia and Gambogi – to exploit the opportunity. The second area
of interest for Ferruzzi is the increasing complementarity between
turnkey industrial contracts and turnkey civil works: the opportunity
has arisen with a few large projects in the USSR (through the creation
of two *ad hoc* consortia: Codest and Convolci). Similar opportunities
are being sought in other centrally-planned economies.

The strategies of private groups, well in tune with the national aid
policy which is discovering the effectiveness of integrated development projects, cannot compensate for the lack of Italian consulting
firms. In this respect, Italy lags behind major industrialized countries
and this can become a serious handicap. Especially in the field of civil
infrastructure, the Italian transport or utilities experts need to set up
full-scale engineering firms (as their French counterparts have done
successfully) and play a larger role in consulting with Third-World
governments. This is a real challenge for the country's export system,
which the state-run groups – ENI and IRI – must start addressing.
In the IRI holding company especially, there has been a courageous
move by the engineering giant Italimpianti to diversify into regional
planning. If co-operation is established with the civil sub-holding
Italstat – which has an unmatched competence in planning, building
and operating infrastructure – IRI will become a favoured partner of
developing nations and open the way for further transfer of Italian
technology.

3 Project Execution

Italian construction firms have chosen to maintain their traditional
skills as constructors, remaining faithful to civil-works specialties
without diversifying (as, for instance, the German giant Philipp
Holzmann has done) or emphasizing service roles through heavy

sub-contracting (as in the Anglo-Saxon version of project management, or the example set by the French Bouygues at the University of Riyadh). A similar attitude still prevails among non-civil contractors who generally base their international success on specialized industrial know-how and are seen as vehicles for exporting machinery 'made in Italy'. The origin of Italian constructors often lies in the discovery by manufacturers that it is more profitable to sell factories than products: an early example was the birth of Fiatengineering from the experience of building car-manufacturing plants in Togliattigrad (USSR) and in Belo Horizonte (Brazil). More recent cases are Danieli (for small steelworks), Cogolo (tanneries) or Inglen (glass plants). Others are SIV, the glass giant belonging to the state-run group EFIM, which is building a turnkey factory in Spain, and SNIA–BPD, a chemical manufacturer belonging to the Fiat Group. In general, even the largest non-civil contractors, while trying to diversify, originate from specialization in given technologies (e.g., Saipem for petroleum pipelines and drilling, with special emphasis on offshore work; Italimpianti in metallurgy). Engineering firms acting as project managers in various fields are less common: the main example is Snamprogetti (although born as the engineer of the oil giant ENI); others tend to have foreign origins, such as Foster Wheeler Italiana, Techint, Technipetrol. Except for these last cases, project execution is thus handled directly by the parent company (local branches acting typically as marketing and logistics supports). Sade/Sadelmi is the only true multinational among the major contractors (fully owned by Asea-Brown Boveri) which has fully-fledged affiliates in each country.

Labour
Italian labour overseas now consists – at least in the case of large contractors – almost exclusively of skilled workers, technicians and project engineers. They tend to be a breed of their own: people who develop a vocation for working abroad and can thus easily be moved from one project site to another, but their fidelity to one employer is quite high. Recruitment of this labour is done nationally (they often join as workers and then move up the ladder through experience overseas). They are employed according to a national contract (the formula being 'temporarily detached abroad'), remunerated according to Italian laws plus an agreed extra sum for working abroad, in a variable mix of currencies. Training is the least formalized aspect as generally 'on the job' solutions prevail, leading to an individual's attachment to a given firm (with whose know-how they identify).

The use of foreign labour is extensive: Italian contractors seem flexible in recruiting from all Third-World countries as Italian project managers are especially appreciated for the ability to adapt and blend with all cultures, the secret always being having Italians as team-leaders. One peculiarity is the wide use of technicians from other industrialized countries (or even people trained in former colonies like India), this is because the Italians lack expertise in foreign engineering and need to adapt to the various regulations.

Equipment and Materials

Construction equipment is procured in a variety of ways (depending on the rules of the host country) and is a typical item to be negotiated as the 'local portion' of the contract (especially if it is left behind and benefits the local economy). The role of Italian contractors as promoters of national equipment is emphasized by the group Fiat: if Fiatallis has grown to become the third such company in the world (after Caterpillar and Komatsu) it is due to the fact that it has been tried and tested in projects the world over.

Links with building-materials industries are the most difficult to assess, especially as Italy has no system of 'trading companies' and building merchants are weak at the national level. In civil works a clear distinction should be made between infrastructure (and large projects in general) and buildings as such (bearing in mind that Italian contractors are much stronger in the former area). While in civil works much use is made of local (or locally acquired) materials, buildings are often erected by prefabrication firms (the two in the 'top ten' list are Sicel and Imprefeal) with their own systems and thus recourse to imported materials (especially for finishings). This is also true of contractors constructing buildings with significant added value (hotels, hospitals, military facilities) whose advantage is typically the superior quality of certain products shipped from home. As for attitudes and policies toward local production, Italy has no tradition either of selling know-how for building systems (and their plants) or consulting; it therefore has no preference or bias, nor does it have a definite policy of fostering local techniques (nor has the government's foreign assistance policy addressed the issue consistently).

Subcontracting and Technology

Attitudes and policies towards subcontracting are not easy to summarize and differ widely depending on the type of project. Industrial plantmakers, for example, subcontract all civil works and, when the main contract is turnkey (as has been the case with all Soviet deals) preferably to Italian ad hoc consortia. Other non-civil contractors (for instance in specialized fields such as pipelines or off-shore platforms) provide their know-how and facilities to third parties in what could be called a form of 'franchising'. Traditional construction firms subcontract all specialized works but avoid labour-only subcontracts, as they prefer to manage manpower directly for a better grip on quality. Subcontracting all sections of work is not typical, as Italian builders are not accustomed to perform as construction managers and, when projects are large, prefer joint-venture formulas, where tasks are clearly assigned.

The involvement of local forces in international projects is of great concern to Italian contractors, although most of their activities are in highly specialized works requiring technical knowledge which cannot easily be transferred. Housing and other simple buildings have never been a market sought after by builders from Italy (with the exception of large and quick reconstruction programmes, like the one following the El Asnam earthquake in Algeria, where Italian prefabrication firms won a quarter of all awards). The only significant exception is the construction of roads, a sector in which pressures for the participation of local builders are mounting everywhere. While smaller Italian contractors are hard-pressed by such competition, the largest firms have managed to create local subsidiaries (Borini & Prono in Nigeria, as an example) which are seen as beneficial for the host country's capacities.

The issue of 'appropriate technology' as such is seen as quite elusive, as it typically applies to the lower end of building activity (generally labour intensive) which is of little interest to large contractors. The answer can be found in 'appropriate development' (as already discussed for the business strategy of Fiatimpresit): it means guiding other countries toward large-scale projects, taking into account all local resources (not just in building) and acting to optimize their mix. Project execution can thereby become a meaningful laboratory for the transfer of technical knowledge in civil and non-civil works alike.

4   Government Policy

Until the mid-1970s it was widely felt among Italian overseas contractors that the availability of stronger financial backing from governments and eximbanks favoured their competitors. They therefore made a big effort to promote their image (large projects abroad were then first becoming a big business) and to convince national opinion that could be a 'foot in the door' for Italian exports of all kinds. They proved to the politicians that the lack of a competitive export credit programme prevented them from increasing their share of a booming international market. The turning point in government policy came in 1977, when the Italian parliament approved a comprehensive and articulated reform (known as the Ossola law) aimed at providing Italian exporters with financial support comparable with that of other major industrialized nations. This law was supplemented by another in 1981 supporting all necessary commercial efforts before the actual bids and extending a global insurance scheme to cover all risks. These laws are deemed by the industry to be among the best internationally. Also created was a healthy climate of competition for foreign lending in the Italian banking community.

Finance
Since 1977 Italian contractors have been able to offer their customers deferred terms of payment at OECD 'consensus' rates by taking advantage of an interest subsidy granted by Mediocredito Centrale. (See the section on 'The OECD Arrangement' in the concluding chapter.) If and when the credit standing of the importer's country is unsatisfactory, the contract can benefit from the insurance cover extended by SACE, the government export credit insurer. Within this system commercial banks and private financial institutions play a crucial role, as official agencies do not lend on their own but simply make up the difference between the OECD 'consensus' and the rate applied to free market lending. Both Italian and foreign banks are allowed to participate in this scheme: the involvement of foreign banks has actually been encouraged by the authorities because of the traditional domestic shortage of financial resources and the higher level of lira interest rates. Coupled with specific Italian know-how, the openness of its export credit market to foreign commercial and merchant banks with more expertise in project financing explains the success of overseas contracting. The tool (introduced by the Ossola law) which has aroused attention in international circles, is the so-called 'triangular policy', i.e., the SACE insurance guarantee which covers loans extended by foreigners in favour of importers of Italian goods and services, and thus successfully steers international banks into lending to higher risk borrowers, as most customers now are.

Public Inefficiency
The main complaints of Italian contractors concern the inefficiency of public administration (which is typical of the peninsula and in utter contrast with the vitality of the private sector): 90 per cent of the industry leaders, interviewed in 1982, identified SACE as the primary cause of reduced competitiveness due to its delays and bureaucratic attitude. Things have somewhat improved in recent years, but contractors claim that a feeling of uncertainty about their government's response penalizes them when preparing the bids in international joint ventures (while their competitors know for sure what they can offer).

On the financial side, Italians would like to see a more aggressive use of 'mixed credits' (as done by the French) – a combination of grants, development aid and export credits well below the 'consensus' rates – and feel inferior to the Japanese, who can enjoy the cheap yen financing raised in their domestic market. (The winning move of joining with them on the Bosporus project will be difficult to repeat without losing autonomy.)

In practice, in spite of the inefficiency of its bureaucracy, Italy has a definite commitment to the promotion and support of contracting business abroad, and foreign policy has become more effective thanks to unprecedented government stability (since mid-1983). The weakest point in this policy is control: insurance being optional (and, according to a Swiss study of 1983, the second most expensive in the world). It only covers the most risky cases and serves what can be called the 'club of major contractors', leaving smaller firms out in the cold. This insufficient control can cause occasional defaults which are dangerous for the image of the Italian industry as a whole.

Foreign Aid and the Italian Mission
Foreign aid policy is quite new on the Italian scene. It was introduced by a law of 1979 but has actually only functioned since 1981. First it supported the backward national design and consulting firms, their contracts having accounted for only 10 per cent of all disbursements between 1981 and 1984. As the appropriations have grown fast to some $3 billion a year, the best use of the funds is seen as being through integrated development projects in which contractors can play a leading role. Aid funds are more and more co-ordinated with export credits in the form of 'mixed credits', and co-financing with international agencies (following the path shown by the World Bank since 1982) is being successfully implemented.

As far as policy at top government levels is concerned, Italy – not having had a significant colonial past and bearing no special military commitments – is in the position of protecting its commercial interests in all countries and playing friend to most. The leverage of trade agreements is often used to obtain the award of contracts (although with mixed results, for example when falling methane values forced renegotiation of an Algerian contract).

Italian contractors, grouped in many associations, find it easy to lobby for their interests at the government level. The high esteem in which they are held can probably be explained psychologically: since her unification in 1870 Italy has failed to become a world power but has had a world presence through the massive emigration of her people. This outflow stopped in the 1960s when the country became a major industrial economy, but continued through the export of construction, of large works that affirm symbolically the Italian international presence. This might explain why, beyond a purely economic calculation of benefits, the Italian government stands sympathetically behind the overseas contractors.

FRANCE 4420
4210
6340

5

France

MARC COLOMBARD-PROUT

1 Trends, Market Shares, and Employment

A few years ago France was third in world construction exports,
behind the USA and the Republic of Korea. This privileged position
was progressively lost to foreign competitors while demand changed
in nature and in geographical location. The decline of new orders
recorded by French contractors followed (with a one year delay, 1983
being the year of large Algerian contracts) the decline of all international contracting from US $143.2 billion to US $64.5 billion (1982
dollars) between 1981 and 1986. In 1985 and 1986 France found itself
in fourth position behind the USA, Japan, and Italy, according to
ENR. But one should emphasize a decline of its market share even in
its traditional export zone in Africa. (Note that French data make
1985 contracts awarded only 77 per cent as high as *ENR* data.)

New orders have shown a significant fall since 1984 in comparison
to previous years. During the 1981–3 period the annual volume of new
orders varied between 60 and 70 billion FF. In 1984 new orders dropped to 45.1 billion FF, a level far below the pre-1984 performance.
Furthermore the gap between contracts 'signed' and contracts
'ordered' (contracts for which financing has been put in place) rose
because of the financial difficulties of clients.

The value of overseas turnover in 1985 was $6.1 billion (1982
dollars) against $6.9 billion in 1984 and $7.6 billion in 1983. Overseas
building work has increased slightly (9 per cent) between 1983 and
1985. Thus the recession is mainly due to civil engineering contractors
who lost 30 per cent of their work abroad (in 1982 dollars).

The 1984 recession was the first since 1979. Great concern resulted
from the simultaneous decline of work on the overseas and the
national market. But one should not forget that the 1984 value of

Table 5.1   *New Overseas Contract Awards, 1982–5*
(US $millions, 1982 dollars)

Year	Building	Civil engineering	Total
1982	2,736	4,728	7,464
1983	2,803	4,167	6,969
1984	1,863	2,910	4,773
1985	1,424	3,187	4,611

Sources: DAEEF, Direction des Affaires Economiques et des Etudes financières/ Direction of Economic Affairs and Financial Studies; FNTP, Fédération Nationale des Travaux Publics/National Federation of Public Contractors; FNB, Fédération Nationale du Bâtiment/National Federation of Building Contractors.

overseas work, in constant French francs, was almost double that of 1974. Looked at from a ten-year perspective, 1985 was one of the worst years, comparable to 1980 but still 10 per cent better than the disaster of 1979.

Number of Firms Involved, Division Among Major and Minor Firms
During the 1977–84 period some 400 building and civil-engineering contractors, each with over 50 employees, exported. According to a 1983 survey of the DAEI (Direction of International and Economic Affairs of the Ministry of Construction) 80 per cent of direct construction exports were billed that year by the ten largest French firms (or groups). Since then, mergers have reduced the number of these to eight. This may still seem large compared with other European countries.

The number of exporting civil-engineering contractors has fallen from 132 in 1980 to 110 in 1985 (out of 5,465 such contractors). Civil engineering work abroad is especially concentrated: the six main exporting groups accounted for almost 80 per cent of turnover. At the opposite extreme, 76 small contractors, mostly subcontractors, realized only 5 per cent of the activity.

Contractors work differently overseas according to their size and specialization. Large contractors have a permanent presence while other contractors are in more specialized submarkets and move about. They often take surplus work as subcontractors for the larger firms. Small general contractors have a hard time keeping their share of the market. In 1984, the contractors who stopped their overseas activity were neither diversified nor specialized.

Table 5.2   *Overseas Contracting Turnover, 1973–85*
(US $millions, 1982 dollars)

Year	Building	Civil engineering	Total
1973	1,076	2,963	4,039
1974	1,124	3,036	4,160
1975	1,728	4,532	6,260
1976	1,774	5,777	7,551
1977	2,254	6,552	8,806
1978	2,744	6,554	9,298
1979	2,082	5,950	8,032
1980	2,198	6,733	8,931
1981	2,049	5,895	7,944
1982	2,143	5,747	7,890
1983	2,058	5,581	7,639
1984	2,593	4,371	6,964
1985	2,241	3,944	6,185

Sources: See Table 5.1.

Share of Overseas Work in French Construction

The share of overseas construction work in total construction output is defined in France as the ratio of work done abroad by mother companies and foreign subsidiaries to work done by the branch both nationally and abroad. During 1967–70 the share of overseas work was 15 per cent. From 1974 to 1977 this ratio was constantly rising: both markets were expanding in value but only overseas work showed real dynamism.

In 1978 and 1979 the national market seemed to compensate for a decline of foreign activity, and the share of overseas work fell by 5 per cent. After a flat 1980, the share rose until 1983 with an annual growth rate of over 20 per cent for foreign work. The peak year was 1983 with a share of 34.8 per cent. In 1985 the share of overseas work dropped to 30.2 per cent, the average for the past ten years.

Employment Generation

Following the Iran–Iraq war, which was extremely costly for the insurance cover of construction exports, public officials began to question the benefit for the French economy of public support for the sector's exports. Construction professionals, through the Union of International French Constructors (SEFI), asked the French associate of Price Waterhouse to evaluate the impact of construction exports on the French economy, considering that Price Waterhouse had made a similar study in the USA.

Using the macroeconomic model, Propage, of the National Institute of Statistics and Economic Studies, the expert report concluded, as reported by *Le Moniteur*, that French construction exports have a multiplier effect of 1.3 on the French economy. In other words, one billion dollars of construction exports add 1.3 billion dollars to the gross national product and create 4,850 new jobs of which 4,000 are in France.

During the 1982–4 period, construction exports directly created 129,000 jobs, including those of 22,000 expatriates, and 30.3 billion francs (about US $3.9 billion 1982 dollars) of annual remittances. Of this amount, 26.2 billion francs, more than 86 per cent, were used in France to buy goods (almost 35 per cent) and services (more than 35 per cent) and to pay wages. The main beneficiaries of the induced effects of construction exports are the engineering, transportation, steel construction, and metallurgy sectors.

Recent Changes in Overseas Work
Bridges and general civil engineering are still the main type of works exported (a third of civil engineering), though their value has dropped by 17 per cent since the peak year of 1983.

Tunnelling and underground works have continued their growth with an annual rate of 35.6 per cent during the past five years. In 1985 the Cairo subway replaced the completion of peripheral hydroelectrical work in the mountainous regions of Guavio in Colombia and Colbun in Chile as the main project.

Marine and port works have enjoyed an important growth in 1985 (33 per cent), electrical works have also increased (18.2 per cent) due to the maturity of construction sites in Latin America and the very high tension lines in the Arabian peninsula.

The share of railways remains stable. The Jijel–Ramdane line in Algeria has taken the place of the Transgabonese.

The increase of work on networks of transportation of fluid energy (gas and oil) is linked to the Iraq–Saudi Arabian pipeline.

Sewer systems and water purification stations have enjoyed constant growth, with a rate 4 per cent higher than overall overseas works, but recently they have dropped due to the divestment of the foreign subsidiaries of a large contractor in this area.

The important decline of road work is the consequence of a fall in orders for new works in Sub-Saharan Africa and the deterioration of the position of US subsidiaries of French civil engineering contractors.

Most types of work declined in 1984 (special foundations, bridges

Table 5.3 *Types of Overseas Civil Engineering, 1980–85*
(US $millions, 1982 values)

Type of Work	1980	1981	1982	1983	1984	1985
Bridge-civil engineering, metal structures	2,007	1,927	1,929	2,038	1,547	1,340
Special foundations, boring drilling	174	178	223	180	117	109
Tunnelling and underground works	81	67	101	91	106	136
Marine and port works	790	737	782	675	414	520
Road and earth works	1,589	1,294	1,239	1,101	929	660
Railways	109	88	89	128	168	150
Water, sanitation, pipe-lines and related installations	1,007	857	725	666	524	405
Public and industrial electrical works	971	755	658	704	565	629
TOTAL	6,728	5,902	5,746	5,583	4,370	3,949

Sources: See Table 5.1.

and civil engineering, marine and electrical works) due to a fall of orders from the Middle East. Bridges and engineering work in 1984 in North Africa did not compensate for the Middle East recession.

Activities that had prospered traditionally were now absent in Middle East markets. Geographical concentration of French construction exports has fallen. If in 1980, the two main client regions represented 55 per cent of overseas activity, they only made up 43 per cent in 1985. Traditional zones such as Sub-Saharan Africa have lacked dynamism while new zones have developed in Asia and Latin America. The important reduction in the geographical concentration of French overseas work is shown by the share of the leading client country: in 1983 Nigeria with 10.4 per cent, in 1984 Algeria with 8.3 per cent, in 1985 Gabon with 6.4 per cent. The five first client countries represented 30 per cent of orders in 1985 against 38 per cent in 1983. The ten first were 52 per cent against 57 per cent. In 1985 Gabon became the leading client, and the USA rose in one year from fourth client to second. Algeria, no longer the first, become the third. The three traditional markets for French contracting are French-speaking Africa, English-speaking Africa (Nigeria), and the Middle East. The share of America (North and South), Europe and Asia rose from 12 per cent to 26 per cent.

2 Business Strategy

The ten principal French firms have merged into eight: Bouygues took over the top French road building group SCREG, and SGE married SOBEA to become SOGEA. They realized sales of US $14.7 billion in 1985 against $14 billion in 1984. These firms had a general strategy to acquire the necessary technical, commercial and financial capability for international competition.

With the decline of international contracts (46 per cent of sales in 1985 against 51 per cent in 1984) most major building and public works contractors have turned back to the national market to consolidate and broaden their basic positions. The international share declined for seven of eight leading firms from 1982 to 1985, but the absolute amount declined for only four (Table 5.4). Practically all the major firms have implemented rigorous efforts at adaptation, new alliances, obstinate searches for diversification and new markets, inside as well as outside of construction.

The first results of the reorganization were visible in 1985 since most major firms had renewed growth of sales, cash flow or profit. Decline of overseas activity was largely compensated for by the extension of their national market shares through the acquisition of medium and small contractors and thanks to a more sustained activity in certain sub-markets such as road maintenance, electricity, rehabilitation, as well as certain 'President's construction sites'. Only three leading contractors, SCREG, GTM-Entrepose and Dumez had declines for a second consecutive year (*Le Moniteur des Travaux Publics et du Bâtiment*, special edition, 'The 500 largest building and civil works contractors', October 1986).

Overview of Individual Strategies
The international activity of the French major firms can be divided into two categories. A minority increased their percentage of overseas turnover in 1985 because of new or current large contracts: Bouygues' international share has grown from 40 to 47 per cent, SAE's from 58 to 60 per cent and GTM Entrepose's from 33 to 41 per cent. While the

Table 5.4   *New International Contract Awards to Ten Leading
French Construction Firms, 1982 and 1985
(US $millions, 1982 dollars)*

	1982		1985	
Firm	Awards ($millions)	Foreign share (%)	Awards ($millions)	Foreign share (%)
1 Bouygues	560	26.4	1,119	41.0
2 Spie Batignolles	1,070	51.6	1,085	49.6
3 Société Auxiliaire d'Entreprises (SAE)	1,505	53.6	1,042	46.5
4 Grand Travaux de Marseille (GTM)	966	56.4	554	29.8
5 TECHNIP	170	n.r.	537	80.0
6 Campenon Bernard	872	63.5	315	36.4
7 CGEE Alsthom	735	55.0	293	29.7
8 Dumez	1,675	88.3	270	60.3
9 Fives–Cail Babcock	330	75.7	99	57.9
10 Société Generale d'Entreprises (SGE)	1,055	51.1	90	n.r.

Source: ENR. These ten firms are all who reported for four or five years during this period. Thirty-three French firms reported for one to three years.
Note: n.r. = no report for the total.

majority have reduced the share of foreign contracts in their general activity, SOBEA has maintained its overseas work around 30 per cent while SGE has reduced it from 46 to 33 per cent, Campenon Bernard from 30 to 28 per cent, Fougerolle from 33 to 24 per cent, SCREG from 50 to 47 per cent. Even Dumez, who still holds the record for foreign sales in its annual turnover, has reduced it from 87 to 85 per cent.

Are major French companies retreating before foreign competitors or simply readjusting to a more realistic and durable share of the international construction market? Are they pausing before a new phase of expansion so as to reconquer their lost position? The answer can only be considered through their new strategies and the impact of the plans that are actually being implemented.

Various strategies exist. Some firms have methodical and discreet strategies of external growth in the USA and Australia as well as technical diversification into electrical works, urban services and industrial maintenance. Others favour alliances with complementary commercial networks to develop barter trade. European strategies are also emerging with the extension of the EEC to Spain and Portugal and the takeover of local contractors who have the prospect of work linked to the Barcelona Games and transport infrastructure in southern Europe.

The Franco–British consortium associating the companies of France-Manche and the Channel Tunnel Group are an example of new partnership strategies for a macro project mobilizing private finance. France-Manche born in May 1985, with the marriage of Trans-Manche Construction Group (Bouygues, Dumez, Spie Batignolles, SAE, SGE) and three French banks (Credit Lyonnais, BNP, Indosuez). The Channel Tunnel Group is composed of five British major contractors (Balfour Beatty, Costain UK, Tarmac Construction, Taylor Woodrow and Wimpey International) and two banks (National Westminster and Midland). The strategy of two groups (Spie Batignolles and Bouygues) can illustrate some of the new plans toward international markets.

Spie Batignolles

Spie Batignolles, the third largest French firm, is a multi-activity group that masters all the trades necessary for the realization of industrial installations and complex infrastructure. Its activities are organized in four divisions: electricity and nuclear (44 per cent), building and civil engineering (25 per cent), engineering and general contracting (17 per cent), oil and pipelines (4 per cent). The importance of electrical activity doubled between 1981 and 1982 with the purchase of Trindel. They have been diversified with the takeover of Fechoz (first French company for stage equipment and scenery arrangements) and of the Company of Electrical Contractors.

Peripheral activities such as parking management (control of access and counting of time) have been given to a new company Elsydel, including Spie Trindel's American subsidiary (Trindel America). Only a minority stake is being retained. In the field of telephones and communication, Spie Batignolles has acquired Automatic Telephone and Teletechnique (1984) so as to extend its activities in private communication systems.

The takeover of Clecim (1985), specialists in the engineering of turnkey ferrous metal units and equipment, was the main expansion of Spie Batignolles (representing an 18 per cent expansion of annual turnover). It completes the palette of Spie's engineering activities that were essentially promoted by Speichim (fertilizer and chemical plants, agro-industry). This division of Spie built the lamination complex of Cilegon in Indonesia in collaboration with Clecim, while Speichim won a Tunisian contract for a phosphoric acid complex, new orders for phosphate calcination units in Morocco, and the construction of a distillery of ethanol motorfuel from corn in Louisiana (USA). This last contract was a source of losses.

Spie Batignolles changed its strategy towards offshore work by reorganizing its subsidiaries concerned with that work – EMH, Intersub, and Anchor Systems. Spie–Capag absorbed Anchor Systems in its oil drilling and pipeline work, an activity with greater turnover than offshore construction.

For Spie Batignolles's management it is impossible to avoid facing up to the decline of civil engineering work. Its objective is to counterbalance the recession by expanding its national building activity through its subsidiary Citra and by the takeover of sound and healthy contractors. These acquisitions will ultimately be useful overseas.

Overseas, Spie Batignolles has farflung operations that have strengthened over the years. Spie's position in industrialized countries (USA, Australia) is strong and often long-established, and is matched by large contracts in developing countries. Spie's international strategy is to stabilize its share of overseas work around 52 per cent of annual turnover, by diversifying abroad. Spie's markets do not fit the traditional image of French contractors working mainly in French speaking Africa: 40 per cent of its overseas work is realized in the Far East and the Pacific zone (particularly in Indonesia over 25 years, in Malaysia and in Australia over 35 years). Twenty-six per cent of overseas work is in Africa, 14 per cent in the Middle East, 10 per cent in America and 8 per cent in Europe.

The Spie Batignolles Group is maintaining its order book at a record level: future participation in the Euro Tunnel Project, the Turkwell Dam in Kenya, a giant gas pipeline in India in association with Japanese, plus a hope of work on the Port Harcourt Refinery in Nigeria.

Spie Batignolles was the object of a raid from Bouygues during 1986. Bouygues managed to buy discreetly enough shares to hold a blocking minority but the attempt failed. Bouygues finally sold its shares. Was this the failure of a takeover or the success of a financial 'coup'? It is difficult to decide. In the process, Bouygues finally gained 75 million FF which was useful at a time when it was buying the TV network TF 1. To avoid eventual future raiders, the Schneider Group increased its share in Spie Batignolles, consolidating control.

In the future Spie is aiming at North European markets. In the field of electrical works Spie has justified hopes in the USA. In general Spie does not expect to win large contracts but rather expects a relatively important number of medium sized ones.

Bouygues

With the takeover of SCREG in 1986 Bouygues has become the unquestioned leader of building and civil engineering in Europe and perhaps the world. In international contract awards, Bouygues was eleventh with $1.3 billion, but its total of $6.5 billion (1982 dollars) put it ahead of Shimizu and Bechtel who each reported $6.2 billion (*ENR*, 16 July 1987).

Bouygues' first international contract followed the completion of the 50,000 seat Parc des Princes Stadium in Paris, built with an innovative prestressed concrete technique. It consisted in finishing the works of the Montreal Stadium in time for the Montreal Games, since bad management and organization were leading to unpredictable delays. This was followed, in 1974 by the 100,000 seat Teheran Olympic Stadium Complex for the Asian Games. More systematic prospecting for overseas work followed. In December 1974, at the yearly convention for staff, Francis Bouygues declared 'each collaborator must consider, at least once in his life, going to a construction site abroad . . . so as to restore our group's expansion by the conquest of third world markets' (quoted in A. Barbanel and J. Menanteau, *Bouygues, the Modern Empire*, Paris: Ramsay, 1987). The period of expansion of national works (1945–73) had ended.

During the 1975–80 period, Bouygues followed a special strategy. Compared with other groups, Bouygues has always had a head start: precocious development overseas as early as 1973, as against 1978 for most groups, and earlier withdrawal from declining markets such as Iran.

Another characteristic of the group is that its international activity is oriented more towards building than public works. This policy towards new building markets is backed up by its national experience.

Overseas we were the first to build housing units abroad . . . France was well placed in the market, thanks to the model policy that revealed itself a very good tool. Thanks to that policy, in fact, the French were without a rival because they could guarantee a product, a price and a deadline. It was important for countries who did not have very developed structures in that field . . . In 1977 we were three French contractors building in Jeddah: Bouygues, Dumez and SAE; and one Dutch company. The Koreans, who worked with us as subcontractors observed everything and copied us. Thus, two years later, in 1979, they are the ones who got the awards (quoted by E. Campagnac in a remarkably documented work on Bouygues: 'Technical and organizational change in building: history of two groups').

The model policy was a good stake for French contractors. But one must point out that some of the operations were not always adapted to the lifestyles of the populations: the 720 housing units in Jeddah and the 1,440 in Ryadh were still empty in 1987. Another example of foreign contracts won on the basis of French experience is the National School of Public Works of Yamoussoukro in the Ivory Coast, won by Bouygues partially on the basis of the reference of the Ecole Polytechnique of France.

Throughout the period Bouygues followed the fortunes of the oil producers in two ways. First through diversification of activities with the creation of Bouygues Offshore in 1975 and a new product (concrete platforms); second by choosing the client countries and ways of dealing with them. The Middle East contracts depended directly on the mother company while three subsidiaries were created in Africa (Setao Ivory Coast, Seprop Morocco, Bouygues Nigeria). Bouygues was very cautious. 'Works abroad have such a dimension, present such a hazardous character that can only suit dynamic, cautious and methodical men. Our worry above all is to minimize the risks and to maintain a light and adaptable structure' (Nicolas Bouygues, January 1978).

The rapid withdrawal from Iran and the limited losses undergone by Bouygues surprised many people. 'Starting from the moment when demonstrations were held at the bazaar every forty days, according to the Shiite rhythm, we understood it was finished, well before the first dead. Everyone was repatriated during the summer of 1978. We simply left locally one to three people from Bouygues with about ten Iranians. Later the Iranians themselves took over the site. They could do it because we had trained them' (quoted by E. Campagnac). One also must add that because of the nature of the clauses included in the deal, construction sites always had a positive return.

The 1980–86 period has seen a new orientation, 'an active scepticism'. Large international contracts are selected more rigorously. Countries that are investigated are chosen according to four criteria: the importance of their needs, the existence of a real political will in the field of construction, strong financial resources, and healthy competition not distorted by the presence of low labour-cost contractors or government-supported contractors. Greater weight is also given to the use of tested methods, integrating a large part of engineering.

Light commitments are favoured because 'to the current risks of any new job, one must add the risks of a job executed far from our bases and in a country whose evolution is sometimes rather unpredictable'. Temporary organizations for the duration of a contract (Tanzania) are combined with the creation of agencies (Iraq, Saudi Arabia, Venezuela). Independent subsidiaries 'whose first objective is to satisfy the laws of the country and who permit a rapid integration in the local situation' are created such as in Nigeria and the Ivory Coast.

The Bubyan Bridge (Kuwait) showed the capacity of Bouygues to win contracts from such an apparently impregnable firm as Brown & Root, the Texan company. A commercial coup, Bubyan is also a technical exploit with its use of prefabricated light triangular structures and an innovative stressing system saving 40 per cent on concrete and 20 per cent on cost. The organization of this project was also very special with a high-tech circuit for transmission of technical and accounting information.

The University of Riyadh followed the Bubyan contract.

The contract of the century signed 18 April 1981 at Riyadh speaks for itself: 3 years of negotiation, a contract in Arabic, a fabulous amount of US $1.72 billion (raised to US $1.95 later), association with the US firm Blount, transport of the US $343 million downpayment by Concorde to New York . . . All that plus something essential: for Bouygues the construction of King Saud University was a fantastic opportunity to learn a new trade. Bouygues had to build without his own labourers, only with computer programmers, lawyers, engineers and managers in charge of an extremely complex system, in English and with American standards. (Gilles Coville, *Le Nouvel Economiste*, 26 April 1982).

In 1982 Francis Bouygues was chosen as 'manager of the year' in France. He pursued his diversification into engineering, with the takeover of the American HDR, specialized in architectural engineering, urban technical services, transportation and hydraulic resources, and environment.

The 1984–6 period was marked by a reduction of international activity and a methodical geographical relocation in Algeria (nine turnkey hospitals in accordance with co-operation agreements with France). Algeria, with Saudi Arabia and Nigeria, represents 80 per cent of Bouygues's foreign activity. Immediate future developments are also linked to the tunnel under the English Channel. Joint ventures or purchased foreign firms are helpful for entry into new countries. AMREP (third largest builder in the world of offshore structures) opened the way to the North Sea and to the American West Coast, while Technigaz (equipment for processing storage and transport of liquefied gas) opened the road to Asia. The years 1983 and 1984 were also marked with diversification into electricity (ETDE – transport and distribution), water, and gas (Saur management of water distribution networks). Since then Bouygues has diversified outside of construction, into FM Radio, cable TV and TV networks with his successful campaign for the privatization of the TF 1. Could it be that Bouygues will start building TV networks abroad?

3 Government Policy

Construction exports had their biggest success in developing countries with unsatisfied primary needs. But the commercial efforts of contractors are now reoriented towards solvent countries: yesterday North America and South East Asia; today Europe and Japan. Contractors and engineers, insurers and bankers, public bodies ask themselves how to cope with the changes. Can one stop supporting construction exports? Can foreign aid policy be co-ordinated with aid to the sector, with private contractors? How? Is it possible for France, with its traditional presence in the world, not to react to the growing gap between industrialized countries and less industrialized countries?

What can be the roles for bilateral and multilateral aid to finance capital formation in developing countries? French government policy is in a phase of evolution; its definition is not finalized. Changes in public policy are being debated through study missions, with private consultancy services and in the secret councils of the Ministries of Construction and Equipment, Foreign Trade and Commerce, and Finance.

The Costs and Advantages of Construction Exports

A study mission was set up by three Ministers of the French government at the beginning of 1985 to survey the costs and advantages of construction exports. This mission, designated the Dreyfus–Bourdillon mission, has not yet (Spring 1987) published its final conclusions. Thus it is only possible to specify its objectives and some provisional conclusions as reported by Gilbert Dreyfus to the seminar on construction exports organized by the Centre of Studies, Training and Information for Engineers of Construction and Industry (CEIFCI) (*La revue générale des routes et des aérodromes*, no. 633, September 1986, and no. 635, November 1986). The expansion of construction exports has contributed effectively to the efforts to put the balance of payments into equilibrium, but in exchange the state has had to accept considerable and growing expenses such as financial support to exports (development of commercial credit and coverage of risks). The objective of the mission is to analyse the effects of construction exports on the French economy, to compare them with other economic sectors, to make comparisons with France's main competitors without neglecting the role of construction in co-operation and development. The mission's role is to answer the following questions: should one maintain, increase or reduce the financial support to the construction sector for the development of exports? An intersectoral comparison with the pharmaceutical industry and the electrical equipment industry for 1984 provides elements to consider. According to the report:

Exports worth a billion francs will generate 3,000 jobs in construction, 3,300 in pharmaceuticals, and 4,000 in electrical equipment.

Per worker value added will be 200,000 FF in construction, 180,000 in pharmaceuticals, and 140,000 in electrical equipment. Where employment is highest, value added per worker is lowest. Multiplying employment by value added gives the value added per billion francs of exports, amounts that do not vary greatly among the three industries: 620 million for construction, 600 million for pharmaceuticals and 640 million for electrical equipment . . . Since construction had required 30 million francs of subsidies, compared with 21 million for electrical equipment, supporting construction is more costly for given results. These results should be considered as preliminary, and as sensitive to changes in assumptions.

The Bank of France has reported that construction has on balance earned 21 billion francs of foreign exchange, while the current balance for all sectors is negative at 78 billion francs.

Political aspects are also taken in account such as the importance of exports and of strong international competition to maintain a modern construction sector and to favour innovation. Realization of spectacular large works abroad represents a permanent 'show room' of French technology.

Though construction exports have a relatively weak return of value added, it is a better sector than others because it is necessary to have high level personnel abroad that weave extremely close bonds with the foreign country. One should also consider the multiplier effects of construction, though the exact amount of these is difficult to measure.

The interministerial mission has not yet published its final conclusions. Thus it is not possible to affirm that the specific and important contribution of construction exports and by consequence the necessity to maintain instruments of aid and support equivalent to international competition will be accepted.

International Co-operation Agreements: The Example of Algeria
A general agreement concerning Franco–Algerian economic co-operation was signed on 21 June 1982. This followed the Gas Agreement between Sonatrach and Gas of France of February 1982 that projected the delivery to France of 9.15 billion cubic metres of gas per year over 20 years.

The gas deal modified the commercial balance between the two countries. Between 1975 and 1980 it was favourable for France by about 4 billion FF per year. In 1981, French imports were slightly superior to exports. But the gas deal inflated imports to France creating a deficit for France of almost 12 billion FF ($1.8 billion US) in 1982. In 1983 France was the first supplier and the first client of Algeria, sales progressed, reducing the deficit to 4.8 billion FF, thanks particularly to large construction contracts signed in 1982 (4 billion FF, US $0.6 billion) and in 1983 (27 billion FF, US $3.5 billion) for works to be undertaken over several years.

Since both states wished to balance their exchanges at the highest possible level, a series of sectoral agreements followed the general deal concerning urban planning, housing, construction and building materials (30 June 1982), transportation (6 November 1982), agriculture, agro-industry and forestry (11 January 1983).

The sectoral agreement for construction comprises several objectives: to lower costs of construction when they are higher than in France for a lower quality, to increase the local capacities of production from 40,000 housing units per year to 350,000, to develop local industries of building materials, to master the problems of urban development, to increase productivity, to render concrete the notion of transfer of technology and know-how.

Technical co-operation is one of the main aspects of the agreement that provides for training of Algerian personnel for supervision and execution of the most modern methods of organization, development of industrialized construction systems, and promotion and reinforcement of the Algerian production potential. Creation of Franco–Algerian mixed (public–private) companies is encouraged and facilitated. The agreement on housing provides for the construction of 60,000 equivalent housing units (housing and accompanying infrastructure) to be realized during three years with a guarantee of good achievement, training and technical assistance.

The sectoral agreement was rendered concrete in July 1983 with the signing of a first contract for 20,800 housing units to be built by small contractors grouped or in association with French major firms such as Bouygues, Dumez, Fougerolle and SAE. The contract has a value of 9.5 billion FF or 1.5 billion 1983 dollars, 51.6 per cent of which is repatriable (about 5 billion FF). Favourable conditions of credit for Algerians accompany the contract: subsidized interest by the French state and buyers' credit for 85 per cent of financing of the repatriable part. International arbitration is provided for in case of conflict, since Algerian public promoters have signed the contract and not the state.

During the same period another contract of a value of 3.5 billion FF (or US $460 million current) concerning railway lines was signed. A significant part of the value of contract, 1–2 per cent, concerns training: Bouygues, like SAE, has set up a school to train its construction-site workers (crane and engine conductors, iron fitters) chosen from local rural labour. Chantiers Modernes train mid-level supervisory staff and job foremen.

Though France has privileged relations with Algeria, the Algerian government must deal with other countries, particularly Eastern countries. Beyond cultural conditions, financial factors play a grow-ing role. Japanese and others have made a strong opening due to the financial conditions they proposed and the Algerian concern for diver-sification of its trading partners. Policy of state to state agreements concerns not only France but also other countries (Yugoslavia and Brazil). The Algerians have concluded a framework agreement with Italian partners. But the will to balance Franco—Algerian economic relations should favour a new flow of orders in the near future.

Prospects
The general strategy of public policy toward construction exports has recently been expressed by Yves Cousquer, director of International and Economic Affairs of the Ministry of Equipment, Housing, Plan-ning and Transportation.

French policy should reinforce strong points in an innovative way and concentrate on countries with high potential. Designing firms should be fostered along lines already prevalent in Anglo-Saxon coun-tries, firms capable of producing complete, timely, and cost-effective plans. As a matter of routine, labour and materials should be purchased wherever they are most economical. A French strong suit is integrated computer planning of industrialized building systems. What needs to be developed further are feasibility and promotional services before the contract, and guarantees, maintenance support and operational assistance after structures are complete. Research and innovation need to be fostered through government planning and European co-ordination.

Elimination of French weak points is also a target for public policy: to fight against French cultural imperialism and to watch clients' needs and desires more closely, avoiding support for unreasonable projects.

French construction exporters must also extricate themselves from national professional feudalisms and typical boundaries between promoters and contractors, structural work and mechanical work, between architects and engineers.

One of the tasks to be undertaken is the adaptation of the mechanisms of credit insurance without using the repatriable part as the absolute criterion, recognizing the positive role played by sub-sidiaries. The game of economic truth in fixing the insurance premium must be played while the government must keep responsibility for the coverage of political risks.

Last but not least, public bodies have supported reorganizations of the French major firms in construction so that they can acquire the necessary technological, commercial and financial capability to cope with international competition.

6

The Federal Republic of Germany

ERICH GLUCH and JÜRGEN RIEDEL

---

1 The Evolution of German Overseas Contracting

Past Trends and Regional Concentration
German contractors have been engaged in foreign work for over a
hundred years. Various impressive examples of large buildings and
civil engineering works such as the Amsterdam Main Railway Station
(1882), the Baghdad Railway, hotels in St Petersburg (now Leningrad)
and Karlsbad, shipyards, harbours as well as many other surviving industrial and commercial buildings still testify to these pioneering
activities.

Precise and skilful modern technical design as well as high quality
and reliability in execution created a solid basis of confidence in
German capabilities and built high standing in international competitiveness. Although during the Second World War business relationships were cut and German companies lost their agencies and
subsidiaries abroad, the traditional high reputation of German quality
helped to re-establish the former business relationships relatively
easily and quickly in the postwar period (Thode, 1978).

By the early 1950s, German constructors had again won overseas
contracts. Turkey, Egypt and Iraq were among the first countries
where German contractors found a footing. During the immediately
following years success extended to Afghanistan, Pakistan and several
countries in Latin America and South-East Asia, while involvement in
Sub-Saharan African countries increased slowly in the aftermath of
their independence in the early 1960s. Since the mid-1970s, however,
these countries have lost much of their importance as clients of German constructors in relative terms. For, as a result of the upswing of
petro-prices, the expansion of demand for construction services in oil-
exporting countries particularly in the Middle East led to a remarkable

increase of total German overseas contracting. This boom, however, came to an end in the early 1980s and business in 1984 dropped below the 1975 level. This decline was mainly due, on the one hand, to financing and balance of payments problems and on the other, to the penetration of the international construction market place by newcomers from newly industrializing countries such as Greece, India, Pakistan, South Korea and Turkey (Riedel, 1983b). Moreover, many countries particularly in Latin America and South-East Asia have now established their own national construction business thereby reducing imports from Germany and other industrialized countries.

Compared with the still considerable German contracting abroad, foreign firms play only a minor role in the German construction market. Even within the European Community where since the early 1960s efforts have been undertaken to liberalize international contracting, no border crossing activities of any size could be observed, except in areas close to national frontiers. For instance, Dutch firms succeeded in winning orders in the northwestern border area while in southern Germany neighbouring Austrian firms traditionally compete successfully in the Bavarian construction market. Sporadically, subcontracting is also offered by Rumanian, Polish and East German firms mainly because their governments can rigorously fix cheap wage and salary rates. Apart from these few exceptions of minor importance, construction within Europe has remained a rather inward looking (i.e., national market oriented) activity.

Volume of German Overseas Contracting

German official statistics have only registered overseas construction *output* since 1977. But analysis can be extended to cover a longer timespan, i.e., one including earlier years with data collected by the Export Committee of the German Construction Industry Association. Since 1950 this committee has recorded all contracts from abroad including project volume, expected schedule of execution and client country. It is assumed that these statistics cover approximately 95 per cent of all orders from abroad.

Not registered, however, are incoming contracts executed by affiliates of German firms located in foreign countries – mainly in the USA, Canada, and Australia – because these contracts are not supervised by the parent company but handled independently by the affiliates. This *indirect* German overseas contracting, though a prominent characteristic of only a few very large construction multinationals, has already attained a considerable volume: according to a recent survey, foreign affiliates of German construction companies independently handled overseas contracts of nearly US $2 billion in the year 1985.

Table 6.1   *German Overseas Contracting: Number of Orders and of Firms, Value of Orders and of Output in US $ millions Per Year*

	Number of		Value of	
	orders	firms	orders	output
			(million US $^a)	
1950/54	23	n.a.	21	n.a.
1955/59	51	n.a.	38	n.a.
1960/64	87	24	57	n.a.
1965/69	197	30	157	n.a.
1970	230	36	160	n.a.
1971	197	29	201	n.a.
1972	260	32	325	n.a.
1973	268	47	415	n.a.
1974	372	42	1,527	n.a.
1975	258	64	2,258	n.a.
1976	341	71	3,968	n.a.
1977	415	72	2,421	2,746
1978	297	56	3,357	3,471
1979	292	51	2,737	3,448
1980	303	47	3,833	3,787
1981	367	56	4,771	4,361
1982	359	49	3,437	4,437
1983	341	46	1,533	4,824
1984	379	42	2,026	4,305
1985	393	40	1,182	n.a.

Source: Export Committee of the German Construction Business Association. Statistical Office of the Federal Republic of Germany (value of output, only).
Note: In this table and throughout the chapter amounts were converted to 1982 values by using the German GNP deflator and 1982 DM-dollar exchange rate.
^{a} at constant prices 1982.

During the 1950s overseas orders to German firms remained relatively modest: on average they grew from US $21 million per year in the first half of the decade to US $38 million per year in the second (see Table 6.1); whereas the number of new contracts increased from 23 to 51 per year, constituting a decrease of volume per contract. This latter trend continued in the 1960s while the annual contract volume went up during the two five-year periods from US $57 million to US $157 million on average. This rise is partly due to the fact that since the early 1960s overseas contracting more and more became an integral part of German development aid in the context of which smaller projects tended to be offered to the German construction industry. This statement applies particularly to building and civil engineering works in Sub-Saharan Africa and some Asian countries where German firms often entered as co- or sub-contractors. This approach was also pursued in the case of larger projects financed by different donors (multilateral aid) where for political and financial reasons the project volume had to be divided among firms from the respective donor countries.

During the mid 1960s contracting with Arab countries was negatively affected by the exchange of ambassadors between Israel and Germany. Ten Arab states cut their diplomatic links with the Federal Republic of Germany, forcing some German construction multinationals to interrupt their business with these countries (Bollinger, 1983). Nevertheless, total German foreign contracting rose to US $400 million by 1973 and then jumped to US $1,500 million in the following year, thus ending a period of slow growth. This jump was due to the multiplication of petro-prices which enabled the oil-exporting countries to accumulate enormous incomes and reserves in foreign currencies. These resources were transformed rather quickly into an unprecedented expansion of construction demand. By 1976 German overseas contracting attained a volume of approximately US $4,000 million – a value which was exceeded only in 1981 (US $4,771 million) – and during the ten year period from 1974 to 1983 German construction firms received orders from abroad of approximately US $30 billion.

By far the most important contract was the construction of the international airport of Jeddah which was handed over to a German consortium led by Hochtief AG in the year 1974. This example shows the order of magnitude of investments carried out in Saudi Arabia at that time. Following the establishment of the basic infrastructure of the airport, finished by 1981, an additional contract was awarded for adding military installations, and even nowadays German firms gain considerable income from maintenance and repair works. Since 1974 the contract volume rose from US $300 million to US $2,000 million (current prices).

During the early 1980s, in the light of the oil glut and together with overall increasing indebtedness in the Third World, German overseas contracting decreased abruptly. In 1983 the volume of new orders was less than half the level of the preceding year and equal to that of 1974 i.e., ten years earlier.

As already mentioned, since 1977 official data are available on both incoming orders and also on German construction output abroad. While during the 1977–84 period this output still oscillated between US $2,750 million and US $4,800 million per year, because of reduced orders the mid 1980s showed a strong decline.

Size of Companies and Contracts

While only some 30 German enterprises were involved in foreign contracting up to the 1970s, this number grew during the 1970s (1973 =
47) and reached its maximum (72) in 1977 before it steadily diminished
to 40 (1985). Most of those engaged in foreign construction are big
companies for whom the share of foreign work in total turnover in
some cases exceeded 60 per cent during the boom years 1980 and 1981.
On average − that is to say, foreign output as a percentage of total
turnover of enterprises with 20 and more employees − this share was
11 per cent during the period 1977–84 varying between 9.5 per cent
and 12 per cent. Of total foreign output, 93 per cent was produced by
companies with 1,000 and more employees, 5 per cent by firms with
between 200 and 1,000 employees and 2 per cent by smaller enterprises. This size distribution has changed little over recent years. Bearing in mind that no data are available for the 1950s and 1960s, there
is only some evidence that in the 1970s more and more medium-sized
enterprises ventured into the international market. For most of them
the booming demand of the oil-exporting countries was the primary
motivation for getting in. Their experiences vary: some failed and
went bankrupt because of their ventures abroad; others performed
relatively well but withdrew from their foreign involvement when the
demand declined and competition from low labour-cost countries
became stronger.

The average size of contract has changed considerably: it first
declined during the 1950s and 1960s, then went up during the boom
years, and slowed down again in the 1980s. Two-thirds of all contracts
now amount to less than US $4 million. This shows that (1) inter-
national competition obliges German firms to accept smaller contracts, (2) due to increasing indebtedness of Third-World countries the
size of projects tends to be smaller and (3) enterprises specialized in
small high-technology projects have raised their market share.

Geographical Distribution and Type of Work

During the 1950s and 1960s the regional concentration of German
overseas contracting depended on and varied with a few big projects
and finally tended towards a few national markets of Asian and
African developing countries. From 1974 concentration increased
with a small though varied number of countries. During the 1974–83
period, out of a total volume of about US $30 billion, 88.4 per cent
of all orders came from oil-exporting countries, 2.2 per cent from
Europe and the balance of 9.4 per cent from other countries. Among
the first group Saudi Arabia played a key role with a contract volume
of US $14 billion, continuing to be during a whole decade by far the
largest overseas market for German contractors. Almost 50 per cent

of all German overseas contracts from 1974 to 1983 were concluded
with this client; in 1984 and 1985, however, this volume declined
sharply to about US $300 million per year. Other important clients
were Nigeria (US $3,700 million), Iraq (US $3,400 million), Libya (US
$2,100 million) and Iran (US $1,500 million). These five countries
accounted for more than 80 per cent of all German overseas contracts.

This predominance of oil-exporting countries considerably
diminished during 1984–6, together with a fall in total volume. In
response to these shrinking markets German construction companies
have made great efforts to diversify and to open new markets in order
to reduce their dependency on the oil-exporting countries. One major
strategy pursued particularly by the larger companies consists of purchasing shares of construction companies in other industrialized countries such as the USA and Australia. This strategy allows them to
establish a solid basis for competing in these markets (Bollinger,
1984).

With respect to different construction categories, German contractors have been involved more in building than in civil engineering projects. Over the last 35 years, 55 per cent of contract volume was
related to building and 45 per cent to civil engineering only. This share
is strongly affected by the large contribution to building investments
in OPEC countries during the 1970s and 1980s, while in the preceding
two decades German contractors emphasized public works such as
bridges, dams, roads, water and port structures. Within the latter
category one-third went into road construction while in the former
public building played a leading role.

Apart from their involvements in overseas construction investment,
German contractors were remarkably successful during recent years in
acquiring contracts covering construction planning, technical design,
consultancy and performance supervision. Contract volume increased
even more in the field of operation and maintenance of the established
structures. Despite quite high growth rates during the last years, the
level of turnover is still relatively modest in this area. It can be
expected, however, that such maintenance contracts may increase
strongly in the coming years.

Impact of Overseas Contracting on Employment
It has been estimated that during the boom years 1980 and 1981 some
10,000 Germans, mainly well-qualified and experienced foremen,
technicians and engineers, were employed on construction sites
abroad (approximately 2,000 in 1973). In addition about 2,500
Germans administered these projects in the export departments of the
companies' headquarters, while about 100,000 workers were engaged
in the host country or from other (developing) countries. The impact

of overseas contracting on German employment compared with the contract volume was rather small as it made up less than 1 per cent of overall construction employment in Germany even during the best years. This may partly be explained by the use of relatively capital-intensive construction techniques and partly by problems of recruiting well-qualified staff at home to be assigned to foreign sites. In particular medium-sized enterprises which started venturing abroad in 1973–5, when full employment still prevailed in Germany, found it very difficult to deploy their scarce qualified staff away from home. On the other hand, the above figures show the important employment effects on the clients' labour market as well as on the emigration from other countries and the respective alleviation of unemployment problems there.

2   The Strategies of German Firms

General Characteristics of German Firms Engaged Abroad

Typical for firms venturing into overseas markets is their long historical record in construction; such companies were usually founded as long as 50 to 100 years ago and have been active ever since in overseas contracting. The ten largest companies account for almost 90 per cent of the total German overseas construction volume (see Table 6.2).

Another feature of these firms consists of their specialization exclusively in the area of construction. Although in some cases they have existed now for about 100 years, these companies have not diversified their activities into other sectors but have specialized in construction and sometimes in building materials. Furthermore all larger companies have created a number of affiliates or purchased other enterprises but have never themselves become affiliates of other multinationals. Also some smaller enterprises have been active in overseas construction since early in this century in specialized fields such as chimney building and sophisticated steel construction.

Of the three major German construction companies, Philipp Holzmann AG is both the oldest and the largest. In 1984 its value of output reached nearly the same amount as the second and third firms combined, and its overseas share (68 per cent) in output is also the largest. Founded in 1849 at Sprendlingen close to Frankfurt, the company received its first foreign contract in 1882: the building of Amsterdam main railway station. From 1903 until 1917 it was involved in the construction of the 1,200 km long Baghdad railway. Continuing its overseas activities – after interruptions by war – the

Table 6.2   *Output of the Ten Largest German Construction Companies in 1984*

Company	Value of Output[a]: (US $millions) total	overseas[b]	Overseas share (%)
Philipp Holzmann AG	3,004	2,043	68
Hochtief AG	1,805	686	38
Bilfinger & Berger AG	1,260	769	61
Strabag Bau-AG	1,119	403	36
Dywidag	876	272	31
Ed. Züblin AG	527	190	36
Walter-Thosti-B. (WTB)	451	68	15
Wayss & Freytag AG	430	151	35
Heilit & Woerner AG	347	125	36
Held & Francke AG	220	57	26

Source: Annual Reports of the ten construction companies.
 Notes: [a] at constant prices 1982;
[b] values include overseas contracts carried out by German firms' affiliates in USA and Australia (indirect overseas contracting).

company became the leading contractor in Saudi Arabia in the early 1970s and realized a large number of projects in both building and civil engineering. One of the largest projects was the overall urban project of Tabuk in the north-western province, when 15–20,000 people were employed in the late 1970s to set up all types of structures. In the early 1980s the company shifted its priorities by investing in US construction companies and making the Jones group of North Carolina a subsidiary. In 1985 Jones was the fifteenth largest American contractor and won contracts worth 1.43 billion (1982) dollars, including $23 million abroad.

Second largest is Hochtief AG which was founded in 1875 in Frankfurt by the Helfmann brothers and incorporated in 1896. Three years later it received the first foreign contract: the construction of grain elevators and related structures in Genoa. This job, a turnkey contract, was then a rarity and the total rate of 3.15 million Swiss Francs a huge amount. Hence, this project can be called a pioneering venture. Another outstanding project was the construction of a large portion of the Avert Canal in Belgium in 1929. This order was twice the company's annual value of output at that time. Its largest contract, however, was the construction of the New Jeddah International Airport. When the contract was signed in 1974, its volume amounted to only 400 million dollars, but when the works were terminated ten years later, the company had completed US $3 billion as general contractor. Other major overseas works by Hochtief over the last 20 years have been the relocation of the Abu Simbel temple (Egypt), the first Bosporus Bridge (Turkey), a subway in Hong Kong, the Mossy-Marsh Tunnel of Tasmania (Australia), 23 turnkey hospitals in Peru, the Cabora-Bassa Dam (Mozambique) as well as the Torbela Dam (Pakistan), the Saddam Dam (Iraq), and finally various harbours all over the world.

Bilfinger and Berger AG is the third largest German construction company, the historical record of which is somewhat more complex. One of the partners goes back to Weis and Bernatz Company, founded 1881 in Speyer, which was transformed into Bernatz and Grün (1886), and later into Grün and Bilfinger (1892). On the other hand, Julius Berger created his company in 1890, and it was not until 1975 that the two companies merged.

Major foreign contracting started with the construction of the Hauenstein–Basis tunnel between Basel and Otten for the Swiss Railway Company in 1911. During the following ten years other engagements were taken up particularly in East European countries such as Romania, Bulgaria, Russia and Turkey. Up to the Second World War works were also contracted outside of Europe. Afterwards, with the construction of large pumping stations and two bridges in Egypt in 1950, Berger's overseas contracting resumed. In 1959 the 135 km long Managil Canal was completed in the Sudan, while only three years later the 9 km long bridge crossing the lake Maracaibo in Venezuela was ready for inauguration. One strategic decision with a long-range impact proved to be the active engagement in Nigeria from 1965 onwards. The construction of the first lagoon-bridge in Lagos was followed by a large number of other projects for which the total contract volume rose considerably following the oil-price increase in the early 1970s. In addition to the status of a foreign contractor, the company entered into a joint venture with a Nigerian partner and thus committed itself more firmly in this country. Large projects included the Juner Ring-Road (1975), the Eko Bridge (1971), and in 1977 − after only 18 months of construction − a complete new harbour in Lagos. Through this achievement, the company gained a high international reputation for beating schedules under adverse conditions. In recent years, however, partly as a result of the oil glut and Nigerian financial constraints, Bilfinger and Berger has reduced its focus on this market by diversifying to a larger number of countries. For example, they bought the American firm, Fru-Con, who were awarded contracts worth 510 million (1982) dollars in 1985 (including $72 million abroad).

Major Types of Construction Works

German construction companies do not welcome questions about their wartime work, especially on bases and fortifications abroad; in any case, these were mostly built directly by the military. Nor have German contractors benefited from the preference of German multinational investors setting up overseas manufacturing plants and service facilities. These multinationals turn to the existing domestic construction business in the host country. There seem to be two major reasons for this: first, German foreign investors are mainly involved in economically advanced developing countries where local construction capabilities in factory and office building are fairly good or excellent and likely to be cheaper. More than three-quarters of Third World German direct foreign investment (DFI) goes to a few mainly Latin American countries. Second, German DFI is too small for attracting German overseas contractors. Hence, contractors only move in, once public contracts are tendered and when the market has a longer term potential for a continuous flow of such contracts.

In general, German contractors have focused on highly complex projects contracted by commercial and public clients such as hospitals, airports, television installations, as well as dams, irrigation systems, power stations, and ports. This emphasis and specialization was accentuated even further in the 1970s and was mainly due to the structure of demand initiated by the rich oil-exporting countries, the major clients of German contractors.

In the last few years a further evolution can be observed: German contractors tend to engage more and more in the maintenance of those structures which they had built in the preceding years. This appears to be a quite interesting market because first these sometimes highly sophisticated structures (hospitals, airports, power stations, etc.) need continuous and careful repair and maintenance. Second, public authorities recognize this need more and more, having realized that the most modern and beautiful newly built structure makes no sense and cannot be productive in the long run unless properly maintained. The lessons of operating faults and lack of service which have transformed new structures into ruins after a few years seem to have been learned and to open further scope for a new and growing contracting market (Bollinger, 1985). A second area of increasing interest consists of activities which require sophisticated technical and organizational know-how, i.e., rendering advisory services in management, planning, modern technologies and their adaptation, engineering consultancy and construction supervision. In these two areas German overseas contractors have been gaining experience and income and see a growing market. In construction itself the market has become smaller, and German enterprises have to cope with hard competition from economically advanced developing countries as well as from the rising local construction industry.

Regional Concentration
The determination of regional priorities, i.e., the selection of individual countries for active engagement is a particularly important issue in the construction sector when contractors formulate their overseas strategies. On the one hand, the fragility of Third World economies, their high exposure to sometimes abruptly changing world-wide economic conditions and their often unstable internal policies make it advisable to overseas contractors to diversify regionally as much as possible. In this sense foreign construction may be called a 'footloose industry', which is sustained by the fact that establishing structures always means a limited period of activity on a determined site, leaving aside ongoing service and maintenance contracts. On the other hand, overseas contracting also requires a longer presence in a country for various reasons:

(1) The first engagement in a country is usually combined with high initial start-up costs and risks, which because of competition from inside and outside the national market cannot immediately be compensated by this first contract. Only consecutive projects may pay for this initial investment.

(2) Due to the fact that a typical feature of this sector is that construction moves from one site to another and sometimes in very remote areas with largely different natural and social conditions, the requirements in logistics, management capabilities and experience are very high and hence their long-term acquirement in an individual country may become a trump card in competition.

(3) Learning by doing effects and their positive impact on the evolution of productivity cannot be overestimated particularly in developing countries with their weakness in infrastructural environment. Due to the usually large share of public construction, this applies in particular to business relationships with government authorities as well as to construction legislation and financial procedures. These aspects, together with the fact that even large overseas contractors cannot be intensively present in the long run everywhere, favour the concentration of an individual enterprise's activities in a few countries and their careful selection within a longer-term strategy. Though there may be some doubts whether German contractors have ever developed such a systematic long-term strategy in terms of regional priorities in a

world-wide context instead of short-term reacting to booming markets, a regional concentration of German interests can be observed. The regional emphasis of the two largest German construction multinationals has been placed on Saudi Arabia, while the third focused on Nigeria. In order to reduce the high risk of dependence on a few countries, however, these companies – as already mentioned – have begun to establish affiliates in the USA and Australia apart from joint ventures in other developing countries. In general, the creation of national companies in particular joint ventures – even with minority equity participation – has become a basic strategy to cope with local content requirements determined by governments in the context of bidding for public contracts.

Consultancy and Other Business Linked to Construction

In line with overseas construction German consultancy has also expanded its overseas activities since the early 1970s. Typical of this overseas involvement is, however, that there are hardly any co-operative links beyond informal contacts between the large construction companies on the one side and the large engineering and architectural design consultancies on the other. Only in specific fields and for specialized works have closer contacts been observed because of the rather limited number of suppliers in these subsectors. Moreover hardly any major close contacts or even enterprise interlacing can be observed between construction companies and suppliers of building materials or of construction equipment in connection with overseas activities.

Bidding and Contracting

German overseas contractors cannot escape from the formal practice used in international contractual design and arrangements which prescribe that the type of contract and the individual obligations already are a constituent of the bidding documents in terms of a ready made contract formula and cannot be negotiated any further. This means that the bidder accepts together with the transmission of his offer the contractual modalities fixed unilaterally by the (public) client, which testifies to the latter's strong position.

Only in the case of negotiated contracts does the contractor have the chance to influence design and other arrangements. According to a 1980 survey by the IFO-Institut für Wirtschaftsforschung in Munich there are usually no negotiated contracts in connection with public works. Instead public contracts are awarded following bidding procedures either open to everybody (i.e., registered and prequalified contractors) or to a limited number of firms capable and/or particularly experienced in respective fields; both types play an equally important role in practice. Private clients, on the other hand, usually do not proceed to a generally open bidding but content themselves with restricted bidding or even negotiated contracting (Behring, Gluch, Russig, 1982). German firms clearly prefer contracts which are concluded on the basis of the FIDIC model which is acknowledged as reasonable and fair. According to the survey mentioned above, however, only in one-third of all contracts could the application of this contract model be carried through.

None the less German contractors in practice apparently attach less importance to the type of contract but orient themselves more towards the type of project, i.e., before deciding on participating in bidding and elaborating an offer, they focus on the likelihood of winning the contract at a reasonable price. For smaller and less sophisticated projects this assessment was usually negative in recent years because German contractors became more and more expensive compared with their competitors. But complex and technically sophisticated projects are still highly attractive to German contractors. They are given to general contractors before being sliced and packaged for subcontracting.

Marketing
All discussions of marketing with representatives of large German construction firms clearly pointed to the importance of personal relations. This applies in particular to the Arab countries. Key persons for marketing purposes are previous clients. According to the results of the above mentioned IFO-survey, these previous clients play an especially important role for small and medium contractors active in subcontracting. For, in fact, public servants representing Government clients and managing public contracts may serve as catalysts for subcontractors to establish contact with general contractors.

For acquisition purposes larger German overseas contractors have established – at least in major client countries – affiliates, often in the form of joint ventures, who maintain small offices and/or have specialized local agents on their payroll. Small and medium contractors usually are not in a position to provide the necessary financing for such activities. This is certainly one major reason why many of these enterprises were not able to continue their operations in a country following the termination of a first contract and why on the whole the total number of German construction firms venturing abroad has declined in recent years (Gluch, 1982). In this context, Iraq might be cited as an example. Due to the shrinking construction market at the outbreak of the Iran–Iraq War, German contractors lost their nerve and gave up their presence in this country. That proved to be a mistake. After little more than a year, the contract volume revived despite the continuing war and tenders were invited for major contracts.

German contractors tend to link themselves in the longer term to a few overseas clients with whom they enjoy a very high reputation because of keeping to schedules and the high quality standards of their work. In addition excellent personal contacts and efficient agents are key factors in success.

Enterprise Co-operation
In order to win overseas contracts it has become more and more necessary to co-operate intensively with local partners. Some countries – among which are the major German client countries of Saudi Arabia and Nigeria – have even introduced legislative and bidding policy measures in recent years which force foreign contractors to establish joint ventures or other forms of co-operation with domestic firms as a qualification for bidding. The creation of such joint ventures, however, is subject to government licensing; and in Saudi Arabia recently, following a period of promoting such co-operation, the government has become more restrictive due to a saturation of business.

German partners point out that a crucial problem with this is that they usually have to take over the entire responsibility for successful performance even if they have only a minority share (as they are legally obliged to have). The growing importance of co-operation in international contracting is also manifested in the increasing number of contracts for which the client prescribes a pooling of contractors or where a pooling might raise the chances for winning the contract. Approximately a third of all German overseas contracts are managed through this form of co-operation. In cases of larger development aid projects where various bilateral and/or multilateral donors are involved, the pooling of overseas contractors from different countries may be – at least informally – imposed by donor countries' interests. The project share of individual contractors then is largely determined by the financial share of their government aid and in many cases in addition by the bank financing that the contractor or the donor agency is able to raise (mixed financing).

The North American construction market as viewed by German contractors has various distinct characteristics which advise specific entrepreneurial strategies. The creation of an affiliate or a joint venture is the only way to get a firm and durable footing in this market. North America is also seen, perhaps somewhat optimistically, as a potential base for penetrating the construction markets of Latin American countries. From a similar viewpoint joint ventures have been established in Australia in order to be closer to the Pacific and South-East Asian markets, where construction demand is growing. These ventures have to be seen as a nucleus for longer-term overseas entrepreneurial strategies.

3 Project Execution: Employment, Materials, and Equipment

When it comes to project execution, overseas contracting involves a number of specific issues on the availability and the procurement of productive factors and other inputs. Major concern is attached to employment, construction materials and equipment.

Employment

During the period of expanding overseas construction, German contractors had to face great recruitment problems at home. This did not only apply to newcomers who started venturing abroad in the early and mid-1970s but also to experienced companies well-established in this business. Foremen and technical supervisors to be assigned to foreign construction sites were hard to find. In addition the firms' managements had to realize that only their best people could be sent abroad in order to be able to cope with the particularly challenging tasks and environments in developing countries.

While unskilled and semi-skilled workers are usually recruited locally or through labour supplying firms from other developing countries, German firms supply higher professional categories such as foremen, chief operators, mechanical engineers, and graduate technicians from their own staff. Moreover, supervision of construction and general management is carried out by German engineers and business managers. In staff selection particular emphasis is placed on personality attributes such as the individual's readiness to engage himself, his capacity to endure hard burdens, as well as his capabilities in organization and personnel management.

As mentioned above, high-level professionals are usually recruited from the German contractor's own staff and are temporarily – mostly from two to five years – assigned to overseas construction sites. This means that their jobs are guaranteed and their employment contracts are not affected in principle but only adjusted to the new functions. After foreign assignment is terminated, this staff is immediately reintegrated in the company at home unless sent to another country. By contrast, most skilled workers are recruited from the general German labour market and receive a specific overseas contract for a limited period, usually of two years. They are not entitled to further employment following this period either in Germany or abroad. This group makes up 80 to 90 per cent of German skilled workers employed in overseas contracting. They may work for much longer periods for the German firm, but the contract will have to be renewed once a project is terminated. Only 10–20 per cent belong to the firms' permanent staff and operate on longer or permanent contracts. The preparation of Germans who are sent on overseas mission is handled by firms in different ways: some content themselves to give short advice on rules and particularities of the respective country, others organize comprehensive training programmes before sending their people abroad.

In many cases, companies report various difficulties in connection with the recruitment of local workers in particular in the rich oil-exporting Arab countries. This is not surprising bearing in mind that during the boom years 1980 and 1981 almost 100,000 so-called locals had been employed on German construction sites abroad. While in early years local agents or specialized labour-supply firms engaged workers for German construction sites from the local or other overseas markets without any qualitative choice, German contractors have increasingly trained local workers in specially established construction training centres before employing them on the site. Though this approach is more expensive at first, it has the advantage that in the long run workers recruited in this way achieve higher productivity and are better motivated. Loss and waste decrease considerably while maintenance and repair is improved. Apparently education in specific centres has also rendered training on the site more efficient.

While in most African and Asian countries unskilled construction labour is abundant, in the sparsely populated rich oil-exporting Arab countries unskilled labour continues to be imported from abroad. Local workers are practically non-existent. Countries of origin have been mainly Pakistan, India, Yemen, the Philippines, South Korea, Turkey, Portugal, Egypt and recently Thailand and Sri Lanka. Most larger German contractors operating on a permanent basis now dispose of a permanent skeleton of workers therefrom. In the meantime some workers have acquired higher capabilities, qualifying them to replace German expatriates.

Construction Equipment and Building Materials
Almost all German overseas contractors operate with cranes and trucks produced in Germany, whereas other heavy equipment used mainly in large road construction and other civil engineering work is

usually purchased from the USA and Japan where major manufacturers are located. Principally construction equipment is purchased but not leased. Leasing sometimes takes place between parent and subsidiary companies.

Bids are generally calculated on the assumption that equipment is no longer usable after the termination of the contract. In fact, however, this equipment can still be employed in many cases. For this reason construction firms make efforts to remain in business by obtaining subsequent contracts. In some cases, equipment which has been entirely written off is offered in another contract without any cost estimate in order to be more competitive, i.e., to raise the chance for winning this contract. This applies in particular to sophisticated equipment and specially made elements in shell construction (for instance in port construction).

In some countries like Saudi Arabia second-hand markets have evolved in construction equipment. These markets, however, apply almost exclusively to all types of trucks, whereas other equipment such as used cranes, for instance, can only be sold at a very small portion of the initial price and thus are not traded on second-hand markets.

Construction materials, which are bulky and needed in large quantities, such as sand and gravel, are purchased from nearest location because of their high transport-cost component. These materials are usually available everywhere and can be used with varying standards. Concerning cement and steel, quality criteria play a much more important role in particular with respect to modern and sophisticated building. In African and Asian countries, therefore, high-grade cement has to be imported from Europe and Japan, depending on building legislation and/or the product standards prescribed by the client. On the other hand, in Latin America these types of cement are procured from the USA and Japan. The predominant demand for low and medium grade cement, however, is generally supplied by factories which have been built during the past 10 to 15 years in a large number of developing countries.

The market situation in the area of steel is nearly identical, though the number of steel producing countries outside the industrialized world is less than those for cement. On the other hand, the production of those steel types most used in construction is relatively uncomplicated. For higher grades, Japanese products have succeeded – due to their better geographical location and pricing – in becoming major suppliers at German overseas construction sites.

Back in the 1960s client (developing) countries had already made attempts to press foreign contractors to employ domestic inputs wherever feasible (local content requirements). Now these efforts have been strengthened. Saudi Arabian public clients, for instance, have requested that contractors show that imports were necessary due to the non-existence of adequate domestic production. Products concerned are wall and floor tiles, door elements, window-frames, wall paper, tapestry and various fittings.

Despite such local content measures by a large number of developing countries, German overseas contractors do not seem to be much affected. Though they express some concern about the supply problems of building materials, equipment, and spare parts, these complaints mainly refer to general supply shortcomings, lengthy customs procedures, infringements of deadlines, financing and payment procedures, claims for warranty and damages, etc. But in most cases these constraints did not constitute major obstacles.

4 Government Policy on Overseas Contracting

Germany has a set of policy measures to encourage industrial co-operation with developing countries, particularly foreign direct investment. But there are no policy instruments such as specific fiscal advantages to promote construction enterprises in overseas contracting. The government saw no need to intervene when the annual volume of overseas contracts declined from US $3,400 million (1982) to US $1,200 million in 1985.

The only exception from this general non-interventionist principle occurred in the year 1983 with a large dam project in Iraq, where German contractors were obliged to interrupt their work because the Iraqi government was no longer in a position to pay the foreign currency component of construction progress. In this situation the German contractors succeeded, with the help of their professional association, in convincing the German Federal Government to contribute to a rescheduling of financing so that work could go on. In this hitherto unique situation the government committed itself in a way exceeding the usual and well established German export-credit insurance scheme (Hermes-Kreditversicherung).

This scheme, though covering the different (exported) construction services, leaves a residual risk of 10 to 15 per cent to the contractor, which, if covered by other insurance requires premiums of about 2 to 3 per cent. Since this amount narrows the profit margin it is not surprising that contractors make efforts to economize on this.

Delays in payment, particularly by public clients are a well-known bottleneck that is neither of recent origin nor is confined to overseas contracting. The reasons are manifold. Partly, these delays are due to administrative shortcomings when the release of budget funds takes more time than originally scheduled. In this event contractors, while practically obliged to continue their work, have to advance their services without being paid. In other cases, original calculations prove to be wrong or outdated during the execution period and have to be renegotiated. Similar procedures result from disputes between client and contractor on the fulfilment of agreed performance criteria. Finally and worst, cases were reported by the business community — particularly in Arab countries — when public clients reinterpreted contracts already signed and tried to modify some to the detriment of contractors.

These problems occasionally led to disputes, litigation, and required time-consuming formal and informal settlement procedures and even international arbitration. Losses were suffered by a considerable number of German overseas contractors. Cases were reported from medium-sized contractors where such losses eat up the profits of almost a whole decade, whereas larger companies were sometimes able to compensate with good profits from other projects. But even the large company Beton und Monier Bau employing some 15,000 people went bankrupt. These problems demonstrate that overseas contracting has not only a sunny side but may also be accompanied by considerable obstacles the removal of which need time, experience and a good sense of partnership and co-operation.

Lesser risks are involved in projects financed by German bilateral financial aid programmes, though these aid projects do not in principle constitute a secure and guaranteed easy market for German firms. Tied aid is not formally permitted. In practice, however, together with mixed financing (public and private) aid is tied to German suppliers to a considerable extent. The contractors' ability to raise funds either of their own or from their banks has an important bearing on competitiveness. This capability is usually greater with large contractors and discriminates therefore, against small and medium-sized firms. On the other hand, this tying does not prevent German contractors from purchasing equipment, materials and labour services from abroad or handing subcontracts to foreign firms. Such a strategy is pursued, especially if their foreign subcontractors or suppliers are able to improve the overall project financing by their financial contribution, including support from their government, thereby reducing the financial cost-component and thus increasing the competitiveness *vis-à-vis* other German bidders.

By 1983, when the number of contracts out to tender suddenly declined, German overseas contractors had learned the importance of financing for successful bidding. One significant example in this respect is the previously mentioned case of Iraq. Since 1983, not only have the developing countries been short of funds, but also the rich oil exporters have lost income and face financial constraints. As a result construction programmes have had to be withdrawn or rescheduled and international tendering declined.

German overseas contracting is not helped by specific government measures such as tax and depreciation allowances, and representatives of the business community have been rather reluctant to request such support. Though raising the issue of international discrimination due to strong promotion and fiscal advantages by governments of other countries, they have hardly ever called for counter-measures to place competition on equal terms; instead their claims are rather of a general nature. In fact, it has been recommended: (1) to channel relevant information fast to interested firms (for this purpose contractors have already set up an information system); (2) to improve communication and to strengthen contacts on political levels for instance through better staffing of German embassies; (3) to support German overseas contractors in financial matters and through better guarantee schemes; (4) to further the efforts to establish standardized contract formulas for overseas construction; and − since in practice only every second contract is in line with the formula so far set up − (5) to achieve world-wide agreement on their application.

5 Conclusions and Outlook

Following an enormous business expansion in the 1970s through the early 1980s, German contractors had to face a sharp decline. This decline was due to (1) smaller foreign revenues of OPEC countries, (2) growing indebtedness of developing countries, especially the more advanced, (3) unfavourable changes in the DM/US$ exchange rate, (4) increasing competitiveness of other (newly industrializing) countries (Riedel, 1983b), and (5) the rise of domestic construction capacities in client countries. Actually these factors will continue to affect German overseas contracting. There is in fact little scope for German firms to counterbalance these tendencies through new strategies. The market for infrastructure projects financed by German development aid is limited. Structural changes within this market tend to favour smaller projects with more local content and domestic subcontracting.

Throughout the world during the past ten years a huge capacity for overseas contracting has evolved which is now searching for employment within a smaller and in many cases further shrinking market. German contractors in general have particular problems in remaining internationally competitive and gaining new and follow-up contracts because of their expensive and high-tech bids compared to those from newly industrializing countries. As a consequence, they tend to engage themselves more in industrialized and some newly industrializing countries, thereby shifting from purely foreign contracting to setting up joint ventures.

S. KOREA 4420
4210
6340

7

The Republic of Korea

DAE W. CHANG

In less than a decade, Korean contractors have become a major force in international construction. They have been highly successful abroad, particularly in the Middle East, which contracted for most of their overseas work. Korean contractors captured about 20 per cent of the Middle East market. Overseas contracts averaged 10 billion dollars per year in the early 1980s (see Table 7.1). In 1984, 59 Korean construction firms operated in 33 nations, employing 150,000 Korean workers abroad. Twenty-five of these firms were listed in *Engineering News-Record*'s 'Top 250 International Contractors in 1983' (*ENR*, 23 July 1984).

The Korean construction industry developed rapidly in the 1960s and 1970s as a result of the implementation of economic development plans and export-oriented strategies. The development of the construction industry was also related to major political and economic events, such as the Korean War (and subsequent period of reconstruction), the military revolution of 1961, the war in Vietnam, the oil crisis, and the Middle East construction boom. The evolution of the Korean construction industry can in fact be divided into three stages, each of which will be examined in turn.

1 The Development of the Industry Within Korea

The Infant Industry (1945–64)

The Korean war (1950–53) destroyed more than half the industrial facilities in South Korea. The reconstruction of the nation after the war afforded both opportunities and challenges to the construction industry. With the change of government following the military revolution in 1961, Korea initiated a series of five-year economic

Table 7.1   *Value of Foreign Contracts Awarded to Korean Firms, 1973–86, US $millions (1982)*

Year	Total	Middle East	Share of Middle East
1973	170	40	23.5
1974	260	90	34.6
1975	815	750	92.0
1976	2,500	2,430	97.2
1977	3,520	3,390	96.3
1978	8,145	7,980	98.0
1979	6,350	5,960	93.9
1980	8,260	7,820	94.7
1981	13,680	12,670	92.5
1982	13,380	11,390	85.1
1983	10,440	9,020	86.4
1984	6,500	5,910	90.9
1985	4,300	3,040	70.7
1986	2,300	1,050	45.7

Source: Ministry of Construction, Seoul, Korea and *Engineering News-Record*, 17 July 1986 and 16 July 1987.

development plans. During the first five-year programme, the Korean construction industry grew by 17.4 per cent per year. A whole spectrum of construction projects – highways, multipurpose dams, housing, harbours, subways, industrial complexes, electrical and communication networks – were carried out. Massive foreign loans financed these projects. Because they were monitored and inspected strictly by the foreign agencies who provided the loans, the projects provided excellent learning-by-doing opportunities for Korean contractors.

A turning point for the Korean construction industry came with its participation in military projects ordered by the US Army Corps of Engineers during the 1960s. Such projects amounted to about US $14 million per year. They consisted of strategic roads, bridges, barracks, camps, warehouses, military installations – mostly building and civil engineering works which were not unfamiliar to Korean contractors. However, these works afforded them invaluable experience working with the US Army Corps of Engineers, who later provided foreign work for Koreans in Vietnam, Guam, and Saudi Arabia.

Benefits drawn from these experiences were various. First, Korean contractors gained technological knowledge, project estimation techniques, skills in international contracting, and work experience in accordance with strict international construction specification standards. Second, the US Army projects demanded obligatory use of equipment and mechanized methods, instead of labour intensive methods, which prompted modernization of the Korean construction industry. Third, the 'Buy-American Policy' of the US required Korean firms not only to purchase US made construction materials and equipment but also to form joint-ventures with US firms in order to avoid limitations imposed by the American policy (OCAK, 1984, p. 21). Finally, the profitability of the US Army projects was greater than that of domestic projects, and those Korean firms that participated directly in these projects were the ones who successfully ventured abroad later.

The Developing Industry (1965–72)
Having acquired the basic skills necessary for international contracting, the Korean construction firms prepared themselves for overseas work by joining the International Federation of Asia and West Pacific Contractors Association (IFAWPCA), which was formed in Manila in 1956 (member countries are Korea, Japan, Taiwan, the Philippines, and Australia). Then, they entered the low-technology end of the construction market in South-East Asia, which is close to home and relatively less developed than Korea.

The first overseas construction job by a Korean firm serving as a prime contractor was a World Bank financed two-lane highway project in Thailand in 1965. The 99.7 kilometre contract was secured at US $5.2 million. Hyundai construction company won the bid against 29 firms from 16 countries, including West Germany, Japan, the Netherlands, France, and Italy. Although the highway was successfully built on time, Hyundai reported a financial loss.

The Vietnam war (1965–75) opened up more opportunities for the international expansion of Korean construction firms. Korea enjoyed a special relationship with the military in South Vietnam in the construction of military facilities, hospitals and port projects, for both Korean and US troops. During this period, Korean firms also searched continuously for new construction markets. They began to work for the US military in Thailand and Japan. Activities extended to the Pacific islands, Iwo Jima and Okinawa, in 1967 and to Guam and Alaska in 1968. In 1971 work began in Brunei, Indonesia, Australia, Canada, and Pakistan; in 1972 in Papua New Guinea, Nepal, Singapore, Samoa, and Cambodia (Shim, ed., 1984). Korean contractors expanded their overseas operations mainly through subcontractor status, and projects were limited to civil-engineering works employing standardized labour-intensive technology.

The Global Industry (1973–85)
In 1975, the fall of South Vietnam left many Korean contractors with no jobs, idle equipment and unemployed workers. Thus they concentrated on expanding their market beyond the Asian region. The Middle East construction bonanza was about to begin.

The Middle East construction market has been the largest in history open to international bidding. Korean contractors were ready to contribute their experience to the infrastructure-development projects in the region. First came the Samwhan Corporation with the construction of a 164 kilometre (US $24 million) highway in Saudi Arabia in 1973. It followed two unsuccessful bids on other projects. Another landmark was the award to Hyundai Construction Company in 1976 of the contract for the construction of the Jubail Industrial Harbour against stiff competition from US, British and German firms. Valued at US $1 billion and employing 3,600 workers per day at peak periods, this project was recognized as one of the most complicated harbour and offshore facilities ever undertaken in the world. It opened the door to the extension of Korean construction services to offshore projects. In 1983, another Korean contractor, Dong-Ah Construction, was awarded the biggest construction contract to date, the US $3.3 billion 'Great Man-Made River' project in Libya, which was expected to employ up to 8,000 Korean labourers on 1,900 kilometres of pipes, bringing water from desert oases to Mediterranean cities. This project has been managed by the US firm, Brown and Root.

From 1973 onwards, the operations of Korean contractors in the Middle East expanded rapidly, as can be clearly seen from Table 7.1. Almost entirely as a result of this expansion in the Middle East, total overseas construction activities grew at an annual rate of 54 per cent during 1973 to 1983, making a significant contribution to the economy and stabilizing growth by counterbalancing the loss of foreign exchange due to oil imports. But the heavy dependence on the Middle East market, which was responsible for 94.7 per cent by value of all overseas construction work between 1973 and 1984, gave cause for concern. The decrease in Middle East construction contracts in the 1980s resulted in a fall in the total value of overseas work undertaken by Korean contractors in 1983–6 (see Table 7.1). The need for geographical diversification of overseas construction activities is therefore apparent. Also despite various efforts to increase contracts in technology-intensive projects with high value-added, Korean contractors still remain most competitive in civil-engineering work and types of building work that are relatively labour intensive.

The Korean Construction Industry in the Mid-1980s
Korean construction firms have a relatively short history. A small proportion of the top contractors was founded after Korean independence in 1945, while others began during the period of rapid economic development in the 1960s and 1970s. By 1983, there were 720 construction firms registered with the Ministry of Construction; 494 were in general building and civil-engineering and 226 were

specialist contractors. Only 35 firms of the total were publicly held and listed on the Korean Stock Exchange. Most of these were nevertheless controlled by founding families. Such firms accounted for nearly 90 per cent of domestic and almost all international construction contracts.

Korean firms today are strong and capable. They have a competitive advantage over foreign firms bidding for contracts in many countries – partly as a result of government financing of pre-bid surveys and insuring against cost inflation. The Korean construction market itself is open to international contractors, especially in areas where local contractors are unable to carry out the required services because they lack experience in using specialized technology. For example, 21 foreign contractors recently applied for pre-qualification for a regional electric-power project. The design phase will be carried out by Koreans with technological co-operation from firms in Finland, Denmark and Sweden. Fifteen firms from the US, UK, Japan, West Germany and Sweden, are competing for the boiler installation and engineering work (*Daily Economic Newspaper*, Korea, 25 October 1985).

The number of Korean contractors licensed by government to operate abroad between 1976 and 1983 is listed in Table 7.2. The value of foreign contracts awarded to major Korean companies is shown in Table 7.3. The decline in the number of firms from a peak in 1979 is immediately apparent, as is also the increase in the number of specialist contractors. Since 1980, cut-throat competition and the world economic down-turn have in fact led to significant concentration in the industry as top contractors like Keang-Nam, Kong-Yong, Chin Hung and Sam-Ho lost out to competition (or 'got into financial difficulties') and were taken over by others. Foreign contract orders have become increasingly concentrated among a small number of firms. In 1983, the top two firms accounted for 45 per cent, and the top five firms accounted for 67 per cent of total foreign work. In 1986 the government announced that foreign operating licences would be lifted from 26 contractors as they finish work in progress, leaving only 15 financially sound firms to cope with the shift in markets.

The largest single Korean contractor was Hyundai Engineering and Construction Company Ltd. Its international ranking in 1985 as a contractor was seventh after five US firms and the Japanese, Kumagai Gumi. In 1984 it had been fourth. It usually accounts for 30 per cent to 40 per cent of total Korean foreign contracts. Perhaps its greatest triumph was the winning in 1976 of the $1 billion contract for the Jubail industrial harbour in Saudi Arabia, as detailed above. Hyundai was founded in 1950. It did extensive work for the US Army during and after the Korean war, as well as later in South Vietnam. It has

Table 7.2   *Korean Contractors Licensed to Work Abroad, 1976–83*

Year	1976	1977	1978	1979	1980	1981	1982	1983
General	54	86	86	84	73	63	58	58
Special	19	35	35	48	39	42	39	41
Total	73	121	121	132	112	105	97	99

Source: Overseas Construction Association of Korea.

worked in more than 16 countries, mostly in the Middle East, especially in Saudi Arabia. Hyundai construction is a part of the Hyundai Group, an industrial conglomerate with interests in the following: construction/engineering, general trading, shipbuilding, vehicles and rolling stock, machinery and electrical equipment, steel and metal

Table 7.3   *Foreign Contract Awards to Major Korean Firms*
(US $ millions)

Company	1966–79	1980	1981	1982	1983	1984 (10 months)	Total
1 Hyundai	5,883	1,413	2,106	3,134	1,109	2,173	15,818
2 Dong Ah	1,945	275	480	494	3,616	28	6,810
3 Daelim	2,337	634	840	1,107	408	426	5,752
4 Daewoo	491	636	2,077	533	891	394	5,022
5 Hanyang	1,021	261	467	954	263	281	3,247
6 Hanil	568	279	718	200	733	97	2,595
7 Sanwhan	817	464	110	588	477	90	2,546
8 Miryung	490	319	310	284	140	148	1,691
9 Samsung	225	360	263	530	58	217	1,554
10 Kukie ICC	260	267	445	162	203	55	1,491
11 Yoo One	451	38	428	27	220	20	1,184
12 Lucky	164	70	426	48	293	147	1,148
13 Dongsan	94	264	59	497	31	58	1,003
14 Poonglim	64	75	156	265	104	23	687
15 Jungwoo	125	98	140	133	106	32	634
16 Korea	201	56	140	94	6	53	550
Total (A)	15,136	5,509	9,173	9,055	8,658	4,242	51,732
Total overseas contractors (B)	23,000	8,259	13,681	13,383	10,444	5,124	74,000 (approx)
A/B (%)	67.0	66.7	67.0	67.7	82.9	82.8	70.0

Source: Overseas Construction Association of Korea.

products, construction materials, furniture, shipping services, an engineering college, and financing (insurance, securities, and the Korean-Kuwait Banking Corporation).

Although Hyundai is a well-known competitor in the field of civil engineering and labour-intensive work, it is also capable of handling special projects, which have included a desalination plant, nuclear power plants, a chemical complex, and steel mills. Its international competitiveness stems from vertical integration, both upstream and downstream, and from its ability to handle projects on a full turnkey basis. It even operates its own shipping services to the Middle East, strictly for supplying construction materials and equipment on time. It has a pier and unloading facilities staffed with its own longshoremen in Bahrain. From here construction materials are barged to Jubail, Jeddah, Abu Dabi, Iraq, and Kuwait.

2 Business Strategy: Contracts, Co-operation and Finance

This section is based on reports and opinions expressed by experts involved in Korean overseas construction. It reflects the strategies pursued by major Korean contractors. These firms did well in international competition because during and after the oil crisis the country committed itself to a strategy of survival through construction exports. The factors most critical for Korean success are an abundant supply of cheap and skilled manpower; an efficient labour organization based upon tightly run 'teamwork units'; a high degree of motivation and the ability to deliver to acceptable standards and on time; an emphasis on technological advancement; an interlocking banking service; and government support. As the international construction market became increasingly a buyer's market, factors such as government support, finance, and the possession of specialized technology assumed even greater importance.

Modes of Entry into the International Construction Market
Modes of entry into the international construction industry are various. Korean contractors have in turn pursued the export of construction workers, subcontracting, prime contracting, and participation in joint ventures and consortia.

Until recently, most of the prime contracts were awarded to US, Japanese, and West European companies. Firms from these countries had the advantage of early knowledge about projects through the provision of technical advice and designs. They were frequently awarded the subsequent contracts. This situation left subcontracting and the exportation of workers as the only modes of entry for Koreans during the 1960s and early 1970s. As subcontractors, they have in fact been particularly successful on massive Saudi Arabian civil projects – roads, housing, airfields, and port facilities.

By the end of the 1970s, however, Korean contractors began to operate as prime contractors. They also began to extend beyond basic civil construction into the more technical and lucrative engineering-related fields. Working as a prime contractor provided experience in a broader range of activities such as design, engineering, procurement, and construction management. In addition, Middle Eastern governments became reluctant to pay a premium for the established engineering capability of Western firms for electrification and petrochemical complexes. Korean contractors offered much lower prices for those projects by breaking them down into several sections. The most technologically difficult parts could then be subcontracted out to specialized Western companies.

As Koreans moved to the more sophisticated end of the construction industry, the building of plants and turnkey projects, they realized benefits by participating in joint-ventures and consortia. Koreans needed Western partners to solve technological bottlenecks, while competitors from developed countries needed the co-operation of the Koreans as a strategy for holding on to business they might otherwise lose. Strategic alliances resulted, combining the Western partners' strengths in design, engineering, systems installation, and financing with the advantages enjoyed by the Koreans in manpower, project execution, and construction materials. Korean partners can also be valuable in difficult markets such as Libya, Iraq, and Africa, where Western contractors hesitate to work alone because of past problems, including political risk and harsh working conditions. The number of joint ventures undertaken by Korean partners increased from three in 1981 to eight in 1984.

Production Strategy: Diversification versus Specialization
The variety and evolution of organizational forms can be a crucial factor in competitiveness. One of the critical distinctions among organizational forms is the difference between specialists and generalists. Koreans first pursued a generalist strategy, undertaking the whole range of construction activities in order both to gain experience and to capture a larger share of profit. As they entered the era of the Middle East construction boom, however, there was a strong need to also develop specialist expertise. It was expected that specialization would bring improvements in technology, thus strengthening competitiveness and profitability; and that competition among Korean contractors would decrease as they competed for a wide variety of specialized projects. In fields like electrification,

chemical engineering, and pipelines, they now find themselves ahead of other competitors.

Specialist organizations are less flexible and are easily hit by excess capacity, but generalist organizations can survive over a wide range of environmental conditions without being optimally suited to any single condition. As Hannan and Freeman (1977) argue, generalist organizations are likely to dominate when uncertainty is high, involving environmental changes, but specialist organizations would fit better with stable environmental conditions. In the face of shrinking worldwide contract volume, increasing competition, declining profits, and fluctuations in market demand, the implication is that large general contractors have a better chance to 'ride out' the current adverse environment. In fact, Korean contractors have attempted to become larger general contractors with more vertical integration of production. They diversified by venturing into businesses both related and unrelated to construction. Some firms have branched out into activities such as banking, retailing, international trade, the leisure industry, and furniture production. Others have sought vertical integration through entering building materials and equipment industries, or heavy industries such as ship-building, mechanical engineering and steel fabrication. The strategy of vertical integration is aimed at taking advantage of the strong backward linkage effect in the construction industry. Some of the major firms have succeeded in this strategy, and have in fact become highly diversified industrial groups (e.g. Hyundai, Dae-Woo, Ssang-Young and Samsung).

Bidding Policy
The practice of awarding construction contracts to the lowest bidder has been of great advantage to South Korea, with its abundant, low-cost labour. An optimal bidding strategy might have been devised if there had been enough previous data about the bidding patterns of potential competitors. But, because of lack of such historical data, Korean contractors had to rely on cost estimation for their bidding strategy.

Korean contractors minimize charges for overhead and the risk premium of doing business in a foreign environment. One Korean executive complained of recent cut-throat competition by saying, 'My only chance of winning the deal is to drop all overhead costs. If I put 5 per cent overhead cost on top of my price I would be third or fourth in a bid.'

The percentage of successful bids by Korean contractors abroad declined from 36 per cent in 1980 to 23 per cent in 1984. This decline reflects the recent loss of the Korean competitive edge and fierce international competition. Despite some accusation of underbidding by

Koreans, the difference between the tender price of the lowest and the second lowest bidder is generally less than 10 per cent. Furthermore, Korean contractors are moving away from cut-rate bids for market expansion and are becoming more cautious and selective about the profitability of projects in international bidding. Meanwhile, in South East Asian markets, Koreans claimed that the Japanese were underbidding or 'dumping'. In 17 cases during 1981 to 1982 the winning Japanese bid price was from 20 to 75 per cent below that of the Koreans.

Trans-Organizational Strategy

In a trans-organizational system, groups of organizations have joined together for a common purpose (Cummings, 1984, p. 368). The organizations maintain their separate identities, but employ either some formal or informal collaboration for joint decision-making.

Korean construction firms in the Middle East and their supporting organizations can be regarded as a loosely coupled trans-organizational system. The system includes individual construction firms, industry associations, financial institutions, and various government organizations (the Ministries of Construction, Finance, and Commerce, and the Economic Planning Board). These meet together on a Monthly National Trade Promotion Meeting which serves as a forum where policies and issues involved in construction exports can be discussed, where co-ordination between the private and public sectors are improved, and where conflicting interests among firms are resolved. Also, Korean contractors working in Saudi Arabia meet monthly at the Korean Embassy and make plans to exchange equipment for rent or sale, often to exchange work crews, or even to swap management teams. Within this trans-organizational system, members interact to improve their chance of success in bidding and to keep their reputations and market shares by assisting and taking over fellow construction firms in trouble. Thus Korean firms as a group in the Middle East are a kind of federative organization.

The Financial Status of Korean Construction Firms

Korean contractors are typically under-capitalized. In recent years, a combination of adverse factors caused a serious cash-flow problem. These factors include Saudi Arabia's reduction of advance payment from 20 per cent to 10 per cent; delayed progress payments totalling 2.2 billion US dollars; decreased profitability; and a high debt to equity ratio.

In 1984, this severe liquidity squeeze hit many firms such as: Keang Nam, Sam Ho, Sam Bu, Kong Young, Nam Kwang, Chinhung, Hanshin, Taewha, Daeduck, Seo-il, and Han Yang. All these attributed their financial problems to big payment delays and ridiculously low bids in the Middle East (the uncollected amount from overseas contracts sharply increased from US \$1.6 billion in 1980 to US \$3.5 billion in 1983). These insolvent firms could not generate a positive cash flow because they had to allocate whatever profit they made to pay interest. Unwilling to bail out financially insolvent construction firms, the Korean government asked them to withdraw from the international market to sell off most assets and properties, to merge with a larger company, or to be taken over by its prime banks.

Also, the inadequate funding facility of domestic banks forced many Korean contractors to borrow heavily from overseas. Foreign exposure to loans totalled 3.8 billion US dollars in February 1985. The relatively high cost of these overseas loans − ¾ to ⅞ per cent above the London Interbank Offered Rate, LIBOR − had to be reflected in bid prices, thus affecting Korean competitiveness.

Korean contractors are seriously handicapped, *vis-à-vis* their competitors, at arranging long-term, low-interest credit for clients. The Korean banking system simply cannot afford to provide enough of a financial cushion for the concessionary interest rates and the deferred export credit payment offered by OECD countries.

The growing practice of forcing contractors to accept oil and other commodities and local currencies in project payments is another headache for Korean firms. Some try to sell oil on the spot market at lower prices (10 to 15 per cent) than the market price. Only the largest Korean contractors, who possess a general trading company within a group, are able to manage counter trade which requires sophistication in selling.

Labour Policy

The ability to attract sufficient quantities of skilled expatriate manpower has been a critical factor in the international construction market. The shortage of technical expertise in the Middle East remains critical, despite high salaries offered. Most Western senior and middle level management personnel would not return for more than two or three contracts because of cultural differences and lifestyles, harsh weather conditions, and physical separation from their families. Here, Korean contractors have a decided competitive edge over their Western counterparts. Korean workers are not only skilled and low cost, but also demand less in fringe benefits and amenities. The cost of maintaining a Korean engineer in the Middle East was in fact estimated in 1978 to be much less than half of what it costs for an American (*Business Week*, 29 May 1978, p. 34).

Another advantage for Korean workers in the Middle East is that Arab policy has changed − from an initial search for skilled Arab labour, and acceptance of workers from the Indian subcontinent, with their Moslem connection – to a preference for Far Eastern workers, who tend to stay for only a short time and therefore do not create permanent immigration. The number of Koreans working in the Middle East rose from 6,600 in 1975 to 57,600 in 1978 (*World Construction*, June 1979). Of a total of 162,000 Korean construction workers abroad in 1983, 150,000 were in the Middle East. Twenty per cent worked for foreign contractors, a figure which the Korean government would like to limit in order to maintain the competitive advantage of Korean firms.

As early as 1979, however, Korean contractors began to hire construction labourers from third countries, at about 56 per cent of the cost of Korean workers. Third country labour is estimated to be only about 80 per cent as productive which brings their cost up to 70 per cent of Korean labour. However, some Korean contractors provide on-the-job training for such workers; and they continue to employ them after a contract is finished in order to build up a cheap and efficient alternative labour force. The percentage of foreign labour employed by Korean contractors rose from 7 per cent in 1979 to 27 per cent in 1983. They are mostly employed in unskilled work.

A most serious labour riot broke out in March 1977 at the Jubail port project in Saudi Arabia. The riot, involving about 3,000 workers, originated from general working conditions, from disputes with management, and from wage differentials. Korean contractors were paying 40 per cent less than the wages paid by European employers. The unrest was resolved in 33 hours. The Korean government announced that it would enforce equitable wages for all Korean employees in the Middle East by introducing a 'standard wage contract' system by the Ministry of Labour. Korean contractors had to submit details concerning wage and working conditions for their Korean personnel before signing a contract. Another outcome of this riot was the formation of a joint labour management conference at every construction site to smooth out any disputes and grievances.

Recently, working conditions at construction sites have been satisfactory: accommodation and other basic needs are provided free and workers remit on average over 80 per cent of their earnings abroad. The total remittance in 1982 was US $1.6 billion – which accounted for 7 per cent of total Korean export earnings.

Construction Equipment, Technology and the Role of Consultants
Equipment possessed abroad by 60 Korean contractors in 1985, totalled 21,900 units of heavy equipment and 58,000 other units (*Daily Economic Newspaper*, 30 September 1985, Korea). Although most of this equipment is fully depreciated over the course of a contract, the

recent decrease in civil-engineering work, which utilizes most heavy equipment, creates some concern for the overseas contractors. The rate of heavy equipment usage was below 50 per cent in 1984. Contractors have a number of options for unused equipment: they can bring it home; sell it in South Asia for a fraction of the original cost; or sell it back to producers, such as Caterpillar and Clark. Recent attempts to bring equipment back to Korea have been seriously opposed by domestic equipment leasing companies and producers that have technology licences with Hitachi, Mitsubishi, Caterpillar, etc. Finding an alternative use for unutilized equipment is an important matter for any overseas contractors.

Much Korean technology originated from Japan and the US, but it has been adapted and assimilated and is now considered Korean. Technology is transferred mainly through the employment of technical staff (Korean nationals) in foreign firms. Licences, technological agreements, and technical assistance from foreign firms now play minor roles (Rhee, Ross-Larson and Pursell, 1984, pp. 106–7). Today contractors only import specialized technology that is not locally available. In fact, the government does not allow firms to bring in whatever technology they like, but tries to make them selective, in order to accumulate technological capability. A survey by the Conference Board (1984), based on 508 US firms, reports that US businessmen identified Korea as having highly competitive construction technology (*The Korean Economic Daily Newspaper*, 4 March 1984). However, in shifting to more technology-intensive projects, Korean firms will need support in R and D (both private and governmental) in order to improve their capacity to handle design, plant engineering, and consultancy.

Korean firms obtain half of their overseas construction materials and equipment from the UK which dominates the construction consultancy business in the Middle East. Korean contractors and material and equipment suppliers suffer from the absence of strong Korean consultancy firms. Efforts have been made to correct this situation. An organization, Korean Consultants International was established in 1982. It is a consortium made of 11 engineering consulting firms plus the Korean Overseas Construction Company (a consortium of small and medium sized contractors). This semi-private company aims to facilitate getting foreign work in engineering consultancy and construction management services. It also provides free engineering consultancy to African nations as aid, promotes joint ventures with foreign engineering firms, and makes efforts to improve the construction design field.

3   Government Policy on Overseas Construction

In the international construction business, a firm's competitive advantage can be increased by various forms of government support. For example, financial or tax subsidies for feasibility studies and bid proposals can be a critical factor in winning construction contracts. The South Korean government was quick to recognize the strategic importance of the construction industry in the implementation of its export-led development strategy. The government's assistance has been extended through regulation; the provision of tax privileges; and through financial, institutional and diplomatic support.

Government Regulation of Overseas Construction Activities

In 1976, the Korean government introduced the Overseas Construction Promotion Law. This law regulates and promotes the overseas activities of Korean contractors in a variety of ways. First, a Korean contractor who wishes to work abroad must be licensed through the Ministry of Construction as a prequalification for foreign projects. Secondly, the government controls competition among Korean contractors overseas. All contractors operating abroad must report their financial status, must meet the minimal capital requirement for a project established by the government, and must submit their bidding plans for approval. The government can intervene to avoid what is called 'excessive competition' among Korean firms bidding for the same project. It also authorizes firms to bid in certain designated countries only. The Ministry of Construction determines the above regulations based on the following track records of firms: past performance in the subject country; past performance on similar projects; references from Korean embassies; capability of equipment, manpower and financing; past and proposed usage of Korean-made materials, machinery, and equipment; past foreign exchange remittance history; and company wage level and union structure. Based on the past two years' performance and current work in progress, the government limits the project size to within 50 times the amount paid in capital and retained earnings of a firm. Finally, in an effort to maintain the excellent performance record of Korean contractors and to deliver a guarantee of satisfaction to clients, the government reserves the right to designate other construction firms to complete any fraudulent works done by Korean contractors who initially won projects. This regulation is regarded as unnecessary and troublesome by most Korean contractors.

Under the Overseas Construction Promotion Law, the government also established the Overseas Construction Association of Korea (OCAK). This is a private and non-profit-making industrial associa-

tion which co-ordinates and promotes the international construction
activities of member firms. The association was originally made up of
106 firms but this had fallen to 51 by 1986. It performs a variety of
functions, including the management of the Overseas Construction
Promotion Fund (see below), the gathering of information and
analysis of international construction activities, and the provision of
a training and welfare service to construction workers abroad.

Tax Incentives

The Construction Promotion Law of 1976 provided a 50 per cent corporate tax exemption for contractors operating abroad. The amount
of tax exemption averaged US $37 million from 1978 to 1982. But this
law was abolished in 1981 to avoid unbalanced growth among different industries. The new tax law provides a tax exemption of 2 per
cent of contract revenue to be reserved for 'rationalization of firms
and overseas market development'.

In addition to tax subsidies, 30 per cent additional depreciation on
equipment is allowed for the purpose of speeding up recovery of
capital investments on equipment. On the individual level, construction workers receive personal income tax deductions up to 715 US
dollars for income earned abroad.

Government Financial Support

South Korea is a capital-scarce country, and it is in no position to provide adequate capital to its construction contractors abroad. In the
past, Middle East clients have provided advance payment of 15 to 20
per cent of the value of the contract − which compensated for the
capital shortage of LDC firms such as the Koreans. Unfortunately,
this is no longer the case; advance payments today are generally only
10 per cent of contract value. The Korean government, which controls
its banking industry, has therefore brought banks to play an
important role through the provision of credit and the issue of bid
bonds, performance bonds, and guarantees. The wholly government-
owned banks, like the Korean Exchange Bank and the Export-Import
Bank of Korea, also make low-interest loans and arrange letters of
credit. The interest rate on loans for construction projects fell from
15 per cent in 1980 to 10 per cent in 1984.

The Korean Exchange Bank and five other private banks − Cho-
Heung, Han-Il, Seoul Trust, First Bank and Commerce Bank − provide standby letters of credit facilities for operating funds and for
machinery, equipment and material procurement. Rates are well
below those of foreign banks. These banks together formed joint
guarantee programmes for overseas construction projects exceeding
US $50 million in amount. A contractor's primary bank issues

guarantees based upon receipt of joint guarantees of other banks. Also, the Korean Credit Guarantee Fund gives guarantees up to 10 per cent of the total amount of joint bank guarantees issued on behalf of contractors.

The Overseas Construction Promotion Fund was created in 1980 to raise US $200 million in five years. It will assure payment to domestic banks in exchange for their issuance of guarantees for construction projects. This support decreased the fear of Korean banks of being overextended to construction firms and avoided the inconvenience and the extra financial costs of turning to foreign banks.

For exporting construction machinery, equipment and materials, firms are eligible for foreign currency loans with a maximum seven years maturity. They require the approval of the Ministry of Finance. These loans include the opening of letters of credit and are booked on-shore.

The Korean government also provides insurance for overseas construction projects through the Exim Bank of Korea. The coverage cannot exceed more than 80 per cent of the contract value. The insurance covers losses incurred by specific risks of failure to receive payments for construction materials and performance on overseas contracts. The risks covered are political in nature: war or civil disturbance; foreign exchange restriction on remittance; bankruptcy of the project promoter; delay in payments extending beyond six months; and force majeure beyond the control of both contracting parties. The Exim Bank of Korea also provides insurance to guarantee facilities for bond issuing banks. The banks are protected from losses arising from the unfair calling of bonds. The coverage is 50 per cent of the bond value.

Institutional and Diplomatic Support
Because project promoters in the Middle East and LDCs are mostly government bodies, the Korean government makes its presence felt in every aspect of overseas construction work. The main government agencies directly involved with supervising overseas construction activities are the following: the Overseas Co-operation Bureau of the Construction Ministry, the Bureau of Trade Promotion of the Ministry of Commerce and Industry, the Office of Middle East Economic Co-operation of the Economic Planning Board, the Bureau of International Finance of the Ministry of Finance, Korean Embassies, the Korean Overseas Development Corporation, the Korea Trade-Promotion Corporation, and the Korea Foundation for Middle East Studies.

Government to government dealings have in fact become very important in the international construction business. South Korea, however, never had the political and historical ties that gave competitive advantages to the UK, France, Italy, and the US, in the Middle East region. The Korean contractors' entry into the region opened a new dimension of political links.

South Korea was able to broaden its international support in the region through the anti-communist policy pursued by most Middle East countries. The number of Korean embassies increased to 15 in 1981 from only 6 in 1975. During this period, the Korean government established the Korea–Saudi Cooperation Committee for Economic Development and invited various governmental officers influential in the construction industry to participate.

It also set up the Korean Overseas Development Corporation – a government agency responsible for placing Korean personnel with foreign firms abroad; it supplies skilled Korean workers for many Western construction firms in the Middle East. And in response to the rapidly growing trade relations between Korea and the Middle East, the Korean Foundation for Middle East Studies (later to become the Korea Industry and Economy Institute) was founded in 1975. It originally had three functions to perform: (1) to conduct basic research related to development plans in the oil exporting Middle East countries; (2) to carry out personnel orientation programmes for Koreans working in this region; and (3) to inform local firms of trade opportunities in the region.

Promoting co-operation with developing countries – South–South Co-operation – has been a major policy of Middle Eastern governments and Arab organizations such as the Arab Fund, OPEC Fund, Arab commercial and merchant banks, the Islamic Banks, and Intra-Arab Institutions (Achilli and Khalid, 1985). Their actions have initiated the emergence of financial, economic and human exchanges in South–South relations on a vast scale. For example, billions in contract value have been passed out to Brazil, India, Pakistan, Turkey, Yugoslavia, Taiwan, the Philippines, and South Korea.

The concept of co-operation within the Third World was further expressed in the May 1981 Caracas Action Programme and the so-called Mecca declaration for a New Economic Order (Achilli and Khalid, 1985, ch. 9). As a part of this South–South Co-operation, Korean contractors try to participate in the economic plans of LDCs by joining in project financing with the World Bank, the Asian Development Bank, and other development-oriented international organizations. Korea also offers a training programme for construction engineers and workers from LDCs (*Korea Herald*, 30 December 1983).

4 Conclusion

The massive decline in Middle East construction projects has seriously affected most Korean contractors since they greatly concentrated their work in this region. Projects in the Middle East still accounted for most of the total foreign construction contracts of Korea but fell from a 91 per cent share in 1984 to 46 per cent in 1986. The Korean share of all international construction had fallen from second during 1980–83 to seventh in 1985. In 1986 Koreans had only a third of the 1984 volume and a sixth of that of 1982. The problems facing Korean contractors are both pressing and various: reduced construction orders and amounts; more localization of works; fierce competition; increasing Korean labour costs (about twice as much as third country workers); lack of financing; limited technology; lack of scientific management skills; and delayed payments or payments in oil from the project owners.

Despite all these difficulties, many Korean contractors still regard the Middle East construction market as sizable and important. Nevertheless, in order to diversify their construction market, they are vigorously pursuing work in Singapore, Indonesia, Malaysia, Hong Kong, Australia, Africa, Latin America, Japan, and the United States (probably the largest, most stable, and lucrative market). At the same time, Korean contractors have become more circumspect and selective about entering the US and other construction markets, and they are still trying to explore competitive advantages *vis-à-vis* local contractors.

As one looks at Korea's ten years of foreign construction experience, it is obvious that South Korea enjoyed success in capturing basic civil engineering projects due to low Korean labour costs. But as international construction contracts become more capital and technology-intensive in nature, Korean contractors are being squeezed out by developing countries like Turkey and India, and by developed ones like the US, the UK, West Germany, Italy, France, and especially Japan with its new financial muscle.

Financing seems to be the most immediate problem facing most Korean contractors. Their short-term financial capacity is limited because of under-capitalization. Financing through the international capital markets in Bahrain, London, and Hong Kong with interest costs of 11 to 13 per cent per year is not competitive. As construction projects in developing regions require more and more project financing, Korean companies simply cannot afford to compete against the attractive financial packages offered by Japanese firms in South-East Asia and elsewhere.

In the light of Korea's declining competitive advantages (higher labour costs and less effective government support), the Korean government emphasizes strengthening competitive advantages of the Korean contractors in the fields of research and development, training, engineering ability, and scientific management related to construction.

The support of the Korean government for its contractors overseas has been extensive and strong. But the excessive government support paradoxically seems to have resulted in the overextension and bankruptcy of some contractors. This situation forced the government to implement a plan to restructure the industry in 1984–6. It consisted of mergers and the acquisition of insolvent firms by larger, financially stronger ones.

Having recognized its two critical policy mistakes of almost automatically giving government guarantees for contracts won abroad and of sending too many Korean contractors abroad, the Korean government will consider alternative policy measures in the future. The government will limit the numbers of Korean contractors licensed to work abroad to 15 major ones. It will also allow only one Korean contractor to bid for any project in the Middle East. It will restrict bidding in high risk countries. It will put more emphasis on narrowing the technology gap between Korean contractors and those from developed nations. Finally, it will eliminate 56 insolvent overseas Korean contractors.

Due to the high amount of uncollected payments from Middle East construction projects, totalling US $3.2 billion as of April 1986, it is difficult to judge completely the overall net benefits of the Korean experience in the international construction market. Nevertheless, as a whole, Korea's international construction experience was worthwhile and contributed enormously to her economic progress in the 1970s. It also proved the point that the construction industry can be a major driving force of economic development for any country.

The future strategy for Korean overseas contractors is clear. First, they must pursue market diversification for their own survival. Second, they must consider forming transnational strategic alliances in the form of licensing arrangements, joint-ventures, consortiums, technological co-operation, and perhaps using franchising concepts in construction and subcontracts. These alliances will help them to overcome the technology gap and to provide financial arrangements and complete package deals for project owners.

8

Scandinavia

STEPHEN P. DREWER

Collectively the four Scandinavian countries, Denmark, Finland, Norway and Sweden, have a population of some 20 million who enjoy one of the highest living standards in Western Europe. They have created what is probably one of the most consistently high quality-built environments. This was achieved through the collective efforts of architects, consulting engineers, contractors and governments whose policies were specifically developed to achieve this end. During the 1950s through to the mid-1970s each country invested shares of national product in new construction at levels rarely exceeded in other countries. Major shifts in product and production technologies were introduced, and resulted in the development of a modern, sophisticated construction industry. Against this background the international activities of the Scandinavian contractors should be evaluated.

1 Background

Scandinavian international construction is, to say the least, remarkably varied. Norwegians constructed in Ecuador, the Sudan, Tanzania, and the United States despite their own oil boom and being themselves importers of construction services. Danish contractors have built radar stations for the US Air Force in Greenland, and Finnish contractors, according to Frederick Forsyth, built the offices of the KGB Directorate at Yasyeno, a suburb of Moscow.[1] Whether this latter is true, or simply 'artistic licence', Finnish contractors certainly have been extensively involved in the Soviet Union on an impressive range of building and civil engineering work over the last 30 years.

The international work of the Scandinavian contractors has a long and impressive history. The construction division of the Axel Johnson Group, one of Sweden's largest manufacturing and trading companies, was operating internationally during the 1920s. Using the 'Amiesit' cold asphalt process, they contracted to surface roads in Finland and Lithuania. Svenska Väg is a part of Johnson Construction Company, which is the main construction company in the Axel Johnson Group. Their interest in international work expanded in the 1930s when it was awarded a major road building contract by the Rumanian State Autonomous Road Agency. This contract, which was signed on 19 March 1931, was for 750 kilometres of road at a contract sum of 70 millions Swedish gold kronor. The contract was completed in 1938 and must surely rank as one of the first major international projects by a Scandinavian contractor.[2] Svenska Väg continues to be engaged on major international projects.

Given our interest in the contemporary activities of the Scandinavian contractors in international construction, it is interesting to reflect on the background to this Rumanian project. The licence to use the 'Amiesit' process had been purchased from the patentees in the USA during the 1920s; an early example of technology transfer. Axel Axson Johnson, a well-respected Consul General, was contacted by the Swedish Legation in Bucharest who wanted to know if the Axel Johnson Group were interested in the project. A condition of the contract was that the contractor had to participate in the financing of the project. Consequently Stockholm's Enskilda Bank, headed by Dr Marcus Wallenberg, became involved and played a key role in the evaluation and financing of the project. A joint Rumanian/Svenska Väg project team was formed to manage the work. This is an early example of arranging long-term finance, a mode of operation not unusual today for international projects.

1950–75, A Period of Development

The main period of development of international construction activities by Scandinavian contractors was during the 1970s. There are however many examples of major international projects built by them during the 1950s and 1960s. 'System' or 'industrialized' building methods were allocated a high priority in the social housing programmes of the Scandinavian countries during the 1955–75 period. Consequently they were, in many ways, at the 'head of the field' in this area of construction. Many Danish builders, for example Larssen Nielsen A/S and A. Jespersen A/S, both developed and exported their 'systems' of heavy prefabrication during the 1960s. Swedish exports of 'systems' were also expanded during the 1960s, for example the Albeton and Skarne systems of Skanska Cementgjuteriet. The Scandinavian contractors were involved in a series of technological developments during the 1950s and 1960s, which established them as a major force in international construction.

As with contractors from many other countries, the 1970s were, for the Scandinavian contractors, the halcyon days in international construction. Although data relating to international construction are open to critical examination, various estimates suggest a growth (in current prices) of between 400 per cent and 700 per cent in the international work of the Danish and Swedish contractors from 1975 to 1984; and Finnish international construction, although rather more volatile, increased in volume by an estimated 500 per cent between 1974 and 1982.

Whilst it is true that international work was expanded in line with a decline in their domestic markets, one cannot be sure that this decline was the major factor influencing the development of an international capacity by the Scandinavian contractors. Many of them, for example Skanska and ABV of Sweden, and Christiani & Nielsen, Hoffman & Sönner, and Höjgaard & Schultz of Denmark, are important European contractors by any standards. Contractors of this stature were bound to respond to the growing international market, regardless of the state of domestic demand. Equally the smaller somewhat more specialist contractors, such as Larssen & Nielsen and Jespersen of Denmark and NPL–Conata of Sweden, who have a significant international profile, supply services which inevitably are aimed at the wider international markets. Nevertheless, the relatively late entry of the major Norwegian contractors into the international markets can probably best be explained by the buoyancy of domestic demand due to the growth of Norway's 'oil-based' economy.

2   Sweden, Denmark and Finland Considered Separately, 1975–85

The Period of Expansion and Consolidation
A major problem when attempting to explain the international operations of the Scandinavian contractors is their diversity of experiences and unique national identities. There are of course many similarities between these countries, but it is mainly geographical proximity which invites treatment as a single Scandinavian entity. In order to overcome this problem certain aspects will be considered at the level of specific countries, while others will be related to scale of operations and technological expertise. This should enable a wider presentation of the

Scandinavian experience to be presented without a tedious repetition of qualifying factors. See Table 8.1 for the trend in Swedish and Danish output, Figure 8.1 for Finnish trends, and Table 8.2 for nine leading companies in 1982 and 1985.

Sweden

The Swedish contractors involved in international construction work are in many ways typical of the larger European contractors. The turnovers of Skanska, ABV, BPA, Johnson Construction Company and SIAB, are in excess of $400 million annually. They each cover the whole range of work of the typical large contractor including building, civil engineering, real-estate development, and certain specialist construction activities. As part of a construction sector, where the 12 largest companies control more than 60 per cent of the domestic market, they have experience in a wide range of modern construction technologies and a high level of managerial competence. Internationally, they have been involved in major civil engineering projects: hospitals, hotels and schools. Usually they operate independently, but on occasions they have been involved in consortia. For example Skanska was part of the consortium which constructed the large Majes Irrigation Project in Peru, while ABV is currently engaged on a major power project in Ecuador, in consortium with Astrup Hoyer their Norwegian sister company.

Skanska and ABV are more intensively and extensively involved in international construction than are the other companies. Skanska was working in the OPEC countries of the Middle East from the earliest stages of their construction boom, and has also worked in North Africa, South America, Sri Lanka, Malaysia and the Comecon countries. Skanska's international work increased significantly during the period 1975–84; as a proportion of total revenue, it doubled. In the 1980s international work, on average, accounted for 23 per cent of total revenue and 30 per cent of total employment. Data for 1984 show their main markets as Algeria (25 per cent), Sri Lanka (12 per cent), Libya (12 per cent), Denmark (10 per cent), Indonesia (9 per cent) and Saudi Arabia (8 per cent). Additionally they have a significant volume of work in Canada, USA, and Norway; countries in which they also have subsidiary or associate companies. More than 27 per cent of 'foreign revenues' in 1984 were from contracts in other developed countries and approximately 50 per cent of this was from contracts in the other Scandinavian countries.

ABV is something of a late entrant into the international markets but is now running very fast. In 1985 they were the largest Scandinavian contractor in *ENR*'s top international contractor listings, having received some $290 millions of new international orders; the most

Table 8.1 *International Output of Danish and Swedish Contractors (in millions of 1982 dollars)*

	1975	1976	1977	1978	1979	1980	1981	1982	1983	1984
Danish	121	190	164	172	203	208	367	330	425	308
Swedish	n.a.	n.a.	n.a.	n.a.	1,038	1,061	821	740	734	834

Source: Danmark Statistik; SBEF (Swedish Contractors Association), Stockholm.

Table 8.2 Contracts Awarded to Leading Scandinavian Companies, 1982 and 1985
(in millions of 1982 dollars)

Firm	Foreign		Total	
	1982	1985	1982	1985
1 Skanska, Sweden	485.0	222.0	1,635.0	1,453.0
2 ABV, Amerad Betong Vagforbattringar, Sweden	274.0	259.6	1,073.0	904.2
3 JCC, Johnson Construction Company, Sweden	n.r.	214.2	n.r.	815.7
4 Christiani & Nielsen, Denmark	249.0	113.7	259.0	139.7
5 Cutokumpu Oy, Finland	n.r.	91.0	n.r.	n.r.
6 Finn Stroi Oy, Finland	172.0	49.2	172.0	49.2
7 Haka Group, Finland	n.r.	42.7	n.r.	614.4
8 Perusyhtyma Oy, Finland	n.r.	14.9	n.r.	n.r.
9 Hoeyer-Ellefsen, Norway	45.0	n.r.	180.0	n.r.

Source: Engineering News-Record, 17 July 1986, 21 July 1983.
Note: n.r. means no report.

Figure 8.1   *International Output of Finnish Contractors, 1974–85*

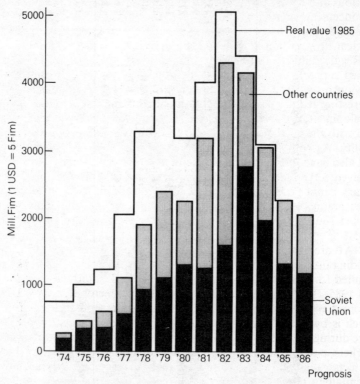

Note: The dark shaded area represents sales to the Soviet Union and the light shaded area contracts elsewhere. The two areas combined give the total volume of Finnish construction abroad in current Finnish marks. The higher line labelled 'Real value 1985' gives the total volume for earlier years recalculated in constant Finnish marks of the lower 1985 value.

important of which was a turnkey contract to build a harbour and the associated infrastructure, on the Pacific coast of Colombia. This international work represented an estimated 30 per cent of new orders received by the company during that year. By 1986 ABV once again lagged behind Skanska with $210 million compared with $326 million in orders.

ABV's involvement in international construction is mainly a feature of the 1980s, growing from less than 10 per cent of turnover in 1980 to nearly 30 per cent in 1984. They also have a significant interest, through subsidiaries, in the other Scandinavian countries and have an operating subsidiary in the USA.

The Johnson Construction Company (JCC) appears to have a more cautious and selective approach to international construction, tending to concentrate on major 'one-off' projects such as the construction of a new port in Rostock (German Democratic Republic), the Olkiluato Nuclear Power Plant in Finland and the INKA sewage treatment plant in Reggio Emilia, Italy. Their international profile appears to be biased more towards the developed countries, although they have previously worked in a number of developing countries. An interesting feature of their international construction activities is the development of markets in Eastern Europe, using their wider association with the Axel Johnson Group to offer clients a counter trading facility. As with Skanska and ABV they also have a US subsidiary, and also have subsidiaries in Scotland and Norway. During 1985 JCC received $217 millions of new orders in the US through their subsidiary Santa Fe Engineers, Lancaster, California. This ranked them tenth among foreign contractors with work in the US. ABV were ranked fourteenth with $105 millions of new orders in the US for 1985.[3]

SIAB are again somewhat more selective, mainly concentrating on building projects in certain targeted countries. International work accounted for less than 2 per cent of turnover and employment in 1984 and 1985. BPA are another major Swedish contractor who have been involved in the international markets. A major construction firm, which is owned by the labour and co-operative organizations, BPA were during the late 1970s and early 1980s, very active in international markets, particularly in North Africa. Problems on a number of these projects gave them cause to rethink their strategy and they are now essentially a domestic contractor. A few of the other larger Swedish contractors have modest international work, but do not seem to consider themselves (as yet) to be international contractors; for example Kullenbergs with a contract in Malaysia and JM–Bygg with contracts in the USA and Norway.

Denmark

Among the major Danish contractors Höjgaard & Schultz, and Christiani & Nielsen appear to be most international. Höjgaard & Schultz has an overseas profile which includes work in Portugal, Angola, Libya, the Peoples Democratic Republic of Yemen, Greenland, and Faroe Islands (Danish) and also work for the US military in a number of countries, including the USA. Their growth in international construction was developed during the 1970s, but they were working in Portugal and Angola building cement factories, harbours, railways, roads and airports during the 1960s. Currently, international projects account for 25–35 per cent of both turnover and employment. A special area of competence, as is the case with many Danish contractors, is work in Arctic areas, obviously founded on the Danish links with Greenland. They have been continuously engaged on Arctic projects since 1952.

Christiani & Nielsen has a long experience of international construction, particularly in the area of civil engineering and infrastructural projects. They are well established in a number of international markets including Great Britain, Brazil and Thailand. As specialists in tunnelling they were working in the Netherlands as early as 1937 and 1942. They also have built immersed tunnels in Canada (1959), Netherlands (1966, 1967, 1968 and 1969), Belgium (1959 and 1982) and the Federal German Republic (1975). In this we have identified the major component of their international success, they are specialists in certain difficult areas of civil engineering construction.

Hoffman & Sönner, a broadly based contracting company, is one of the more important medium sized Danish contractors operating internationally. They have been working in Brazil since the early 1970s and more recently in Saudi Arabia, UAE, and Algeria. In addition they have also worked in the Federal German Republic, and in Greenland and the Faroe Islands since the early 1960s. Their international work tends to be concentrated on engineering services and industrial installations. Monberg & Thorsen are another medium sized Danish contractor involved in international construction work. They have worked as building and civil engineering sub-contractors to the Lummus/Thyssen Reinstahl consortia on the Basrah Petrochemical Complex in Iraq, in consortia with F. L. Smidth, Denmark's internationally renowned producer of cement plants, on a new cement plant in Kufa, Iraq, and as main contractors on four Gas Bottle Filling Plants in Iraq. In addition they have worked in Vietnam with the Danish Sugar Corporation, in Greenland building radar stations for the USAF, and on marine works in Ghana. This is an impressive array of international projects for a company whose annual turnover in 1985 was less than $100 million.

As we move from the larger to the smaller Danish contractors, we notice an increasing tendency to market as specialist subcontractors and in consortia. This latter tendency is an important feature of Danish international construction. Danish contractors have formed consortia for work in Arctic areas (Danish Arctic Contractors), for work in the USA (Danac Inc) and Danfarm for the agriculture and food processing sectors. In addition they are involved in consortia with contractors from Sweden, France, Italy and the Federal German Republic for special projects. It is through consortia that many smaller Danish contractors enter the international markets. E. Pihl & Sön, a company whose annual turnover in 1985 was around $40 millions, is involved in consortia for work in Iceland, Greenland, Yemen, Mali, Somalia, Guinea and Liberia. Other smaller Danish contractors follow a similar strategy.

Finland

When considering Finnish international construction activity, consortia are even more prevalent than in Denmark. Finnish contractors produced $450 millions of work in the international markets in 1985, of which an estimated 60 per cent was in the Soviet Union, 28 per cent in the OPEC countries and the rest mainly in Africa. As was stated earlier, there was a five-fold growth in real terms during the period 1974–82, and the proportion of this in the Soviet Union rarely fell below 50 per cent of the total.

Finn Stroi, a consortium of some 15 Finnish contractors, was established in 1972 to co-ordinate the large-scale turnkey projects of Finnish contractors in the Soviet Union. As an autonomous unit, Finn Stroi accounted for 15 per cent of Finnish contract construction exports in 1984, while the member companies were responsible for 84 per cent of the total output. Their work in the Soviet Union is directly linked to the Finno-Soviet trading agreements and involved a significant element of counter trading. Many of the projects are in the area of pulp and paper processing, in which the Finns have great competence. Typically they are responsible for the supply of a total process, including design, construction, materials and components, and labour. As such these are 'turnkey' projects in probably the ultimate sense. Among their most important projects in the Soviet Union are the construction of the iron ore mining and processing complex in Kostumuksha, the Svetogorsk pulp and paper mill, the Vyborg pulp and paper mill and the Norilsk Dairy. More recently, with the anticipated downturn in their Soviet markets, they have targeted South East Asia and Africa for future expansion.

Among the largest Finnish contractors is the Haka Group, which is a member of the Finn Stroi consortium. Haka is also, however, an international contractor with a separate identity. Either through Finn Stroi or independently they contributed 14 per cent and 15 per cent of total Finnish construction exports to the Soviet Union and other countries respectively in 1985. Haka is part of the Enka Corporation, a conglomerate comprised of 116 individual companies which, with sales of $3 billions in 1985, is the third largest company in Finland. The Enka Corporation, a co-operative with links to the Finnish trade union movement, is a multi-product conglomerate, divided into three main sections: Enka, which is involved in the production, distribution and import/export of a wide range of goods and services; Haka, which is involved in construction and project exports; and Kansa, which concentrates on insurance and non-banking finance for both the domestic and international markets. The Haka group, which accounts for roughly 25 per cent of the Enka Corporation's sales, is itself effectively a construction conglomerate. Construction exports are centred on the 'parent' company, the Haka Construction Co-operative, and three specialist 'subsidiaries': OMP-Yhtymä which specializes in civil engineering and industrial construction, Insion-öörirakentajat and Vessi-Pekka, which specializes in international building and civil engineering construction. Vessi-Pekka has been recently assigned the major responsibility, within the group, for the future expansion in the international markets.

The Finno-Soviet trading relationship should be explained in more detail. The Finnish and Soviet governments have agreed to a series of five-year trading agreements, and the sectoral distribution of this trade is agreed at ministerial level. Consequently a building quota is established, which for the period 1986–91 is 1.2 billion Soviet roubles. This trade is essentially counter trade, which implies a particular set of strategies for the Finnish companies involved. The problem for the Finnish contractors is to accommodate both the complexities of counter trade and the mechanisms for dividing, among themselves, the total building quota. Individual contractors operate within the Soviet Union both as part of Finn Stroi and as independent contractors. The distribution is related to scale and type of projects and the relative construction intensity of the projects. Haka's 14 per cent market share of Finnish construction in the Soviet Union directly reflects their share in the Finn Stroi consortium, which implies a low level of independent activity by Haka in the Soviet market during 1985.

Another Finnish contractor with an extensive international profile is YIT Oy. Their first international contract was in 1958, for a water treatment plant in Kerbala, Iraq. Water and sewage treatment facilities were the original specialist area of the company and they have operated internationally in this field for nearly 30 years with projects in Saudi Arabia, Libya, Abu Dhabi, Tanzania and Kenya. They have also worked on a wider range of projects such as public buildings in Saudi Arabia, hotels in Poland, recreational centres in Iraq and a depository for the Tretiakovsky Gallery in Moscow. In some cases this work has been as an autonomous contractor, sometimes as part of international consortia, at other times as specialist subcontractors or as part of a wider Finnish package. During the early 1950s international activities accounted for more than 50 per cent of company turnover, but this proportion had fallen to 25 per cent by 1984. Exports to the Soviet Union were 10 per cent of total sales in 1984 and 35 per cent of total international activities. At 10 per cent this again is directly comparable to their share in the Finn Stroi consortium. As part of Finn Stroi they worked on the Svetogorsk Pulp and Paper Mill (1985–88). Independently and in separate joint ventures, they have also contracted for food storage facilities, industrial plants and the new port of Tallinn, all in the Soviet Union.

Many other major Finnish contractors are working internationally; for example Puolimatka, part of the Rakennusvalmiste construction related conglomerate and Polar and Partek, also construction related conglomerates, all have a major interest in international construction. They are all extensively engaged on contracts within the Soviet Union and also involved in other international markets; for example the OPEC countries, Africa and South East Asia. Although Finnish international construction is founded on the Soviet markets, it is now international in every sense. While contracts awarded to the leading four Finnish firms rose by 152 per cent during 1985–6, that of the four leading Swedish and Danish firms fell by 17.2 per cent, more than the average international decline of 11.7 per cent. Nevertheless Swedes and Danes still won a third more in contract awards than the Finns.

3   A Wider Perspective

Presenting Scandinavian international construction as the activities of selected contractors from each country is bound to omit much Scandinavian involvement in the international markets. For example Calor-Celsius, the largest Swedish engineering services contractor, has a major international profile with work in Denmark, Norway, Finland, the Federal Republic of Germany and the USA. They have also worked as specialist subcontractors in the Middle East and North Africa. Their electrical engineering subsidiary Vanadis, has a similar range of international activities. Many of the major contractors have engineering services subsidiaries who are used to service the parent company's international projects.

The Scandinavian presence in international construction comes in diverse packages. For example Superfos, Denmark's largest industrial conglomerate, has a road-construction division with contracts in Egypt and a subsidiary in the USA. Similarly NPL-Conata, a medium size Swedish construction company, is extensively involved in erecting their patented building system for clients in a wide range of developing countries. Similarly A. Jespersen & Sön and Larssen-Nielsen build their own patented systems in many diverse countries throughout the world. In every Nordic country there are major producers of prefabricated 'timber-framed' houses, usually subsidiaries of the larger contractors, who have a major share of the international market. In fact this is the most important category of work in the Haka Group's exporting activities.

We should not restrict ourselves to those issues which are 'construction specific' when considering the international activities of Scandinavian contractors. The construction sector in many ways derives its main characteristics from the wider political, social and economic environment which both conditions and stimulates its development. The Scandinavian countries have a number of common characteristics in this context. Politically, they have a common commitment to social welfare, hence the scale of their social housing, educational and health related building programmes. Socially they are relatively cohesive, reflecting a low level of social tensions, which spills over into industrial relations. Economically, they are marked by industrial concentration, vertical integration and a heavy dependence on international trade.

These factors condition Scandinavian construction in a number of ways. Public-sector intervention as a client and manager of the national construction process, particularly in the 1960s, was instrumental in determining the volume of demand, technological developments and, by extension, the structure of the construction industry. Due to the size of projects, industrialized prefabrication techniques, and changes in forms of contract, the industry has become concentrated to a degree not found in most other industrialized countries. A co-operative and relatively quiescent workforce has been a significant factor in the introduction of new technologies, leading to increased efficiency and product performance. The existence of large multinational corporations in certain sectors, such as Alfa-Laval, Ericssons, and Volvo has enabled domestic contractors to benefit from the experiences of working at the frontiers of modern production technology, and provided a link with an international manufacturing network through involvement in the turnkey projects and direct overseas investments initiated by the major manufacturing companies.

There are, however, other factors which have been conditioned by this wider environment. Vertical integration is pervasive and has enveloped construction-related activities, such as the roadbuilding subsidiary of Superfos, or become a model for the construction sector, for example the Finnish companies. Similarly there is what might be termed a technological pull-effect which tends to equalize the level of technologies in other sectors, consequent on the 'pull-effect' of the dominant advanced technology sectors. For example the construction demands of Volvo entail time and quality requirements which imply an enhancement even of 'traditional' construction technologies.

The same factors which were instrumental in stimulating concentration in the construction sector also stimulate the move into international construction. A domestic construction market which services a population of either 5 or 8 million is not sufficient to satisfy a major contractor. He must either internationalize, diversify, or internationalize *and* diversify. Most of the companies mentioned have chosen the latter option; those who have chosen to concentrate solely on diversification, such as Lundberg, one of Sweden's largest contractors, effectively cease to be construction firms.

4   Executing the Projects

One of the crucial determinants of success in international construction markets is the effective execution of projects. Whatever their country of base, logistics, management, and procurement are aggravated when a contractor operates in a new market. When working internationally additional burdens arise. Such 'resourcing' also has major implications for the net export component of a contractor's international operations, which in turn is an important factor in determining the level of support provided to their contractors by national governments. Scandinavian governments offer a complete range of support services to assist their contractors operating internationally. Currently research is underway to examine the extent to which these support services are adequate given the actions of other national governments with respect to their own contractors. I have not listed the various supports available to the Scandinavian contractors due to the limits of space and a wish to avoid being tedious and repetitive.

There is no reason to assume that the resourcing strategies of the Scandinavian contractors are unique. However there are certain aspects of these strategies which may be unusual and as such should be discussed. Generalizations are always subject to qualifications and the following are no exception. The labour supply function of Scandinavian contractors differs according to the broad geographical areas and economic characteristics of client countries. Typically they can be divided into three categories: developed countries, developing countries (broadly defined) and the Soviet Union and Comecon countries.

In other developed countries they will tend to operate in a similar way to the local 'national' contractor; in fact their local subsidiaries will in many ways see themselves as local 'national' contractors. Host country nationals are employed at every level including the technical and management categories, while only at the most senior levels will there tend to be a bias towards Scandinavians. When working in developing countries they tend to staff projects with Scandinavians to the level of production management and use local or other 'nationals' as production workers. There is some evidence that they also employ 'nationals' of other developed or South East Asian countries at senior levels of project management. Conversely for work in the Soviet Union and the Comecon countries, they will typically provide the complete range of labour requirements from their domestic base.

Skanska, for example, extensively used labour from Thailand when working in Saudi Arabia and the UAE in the 1970s. As they moved with new contracts into other areas they moved these same Thai workers into other countries, including Algeria, a country in which there was, in principle, a reasonable supply of local labour. The reason is obvious, having trained and developed their Thai workforce in 'Skanska ways' they were not keen on 'starting from scratch' with a new group of workers. However when working in the Soviet Union on a new hotel in Leningrad their workforce was almost exclusively Swedish. Conversely while building the Kotmale Dam in Sri Lanka, their site workforce was almost exclusively local.

Finnish contractors in 1982 employed for international contracts nearly 11,000 workers, of whom 5,000 were Finns, 2,000 citizens of the local country and 3,800 from other countries. Approximately 25 per cent of the Finns employed were 'salaried' staff and the rest 'wage earners'. However over 90 per cent of wage earning Finns were employed in the Soviet Union where they 'rely almost entirely on Finnish labour'. About 800 Finns were working in the Arab countries, where most of their labour comes from the Philippines, Pakistan, Bangladesh, India, Thailand and Sri Lanka. In other areas such as Sub-Saharan Africa the labour force is drawn mainly from the local populations. In fact less than twenty Finnish 'wage earners' are employed in these other countries. Finnish contractors claim to employ citizens of 25 different nationalities on their international construction projects.

Danish contractors have an employment profile similar to their Finnish and Swedish counterparts. When Danish companies worked elsewhere in Scandinavia, Danes made up 56 per cent of their labour force. Elsewhere the percentages were as follows: other Western Europe, 10; Middle East, 8; Far East, 2; Africa, 33, and other countries, 10. It should be noted that 'Africa' includes Algeria and Libya and 'other countries' includes the USA.

When considering the 'materials' resource, a similar division of markets is apparent. For example in other developed countries the local materials supply system is well developed and can compete effectively with Scandinavian imports. Success on work in developing countries requires of necessity the use of the best 'local' and 'international' suppliers. Contractors could not compete effectively if in allocating priority to their own domestic producers of materials, they neglected to develop competence in international procurement. However when working in the Comecon countries, Scandinavian contractors use their home suppliers for a remarkably wide range of materials and services.

A number of points are relevant to an understanding of the Scandinavian contractors' materials procurement strategies. Most contractors are part of a vertically integrated construction-related conglomerate; at times the parent company. Consequently their international construction activity is not only related to an interest in contract construction, but also to the wider materials-supply factors. As such there are strong internal pressures to maximize the benefits of international operations for the company as a whole. We can extend this argument to include the Finnish consortia where the pressures for maximizing the Finnish content are strong.

There is however some evidence that the Scandinavian preference extends beyond intra-company objectives. Analysis of the material and services inputs for selected international projects of some Swedish contractors shows a major Swedish component. For example one Swedish contractor used domestic suppliers extensively on a major hotel project in Baghdad, placing a total of 1,500 separate contracts with Swedish suppliers, including 81 which were valued at over $100 thousand, and 286 greater than $10 thousand. Another Swedish contractor even imported coffee from Sweden for a project in the German Democratic Republic. Danish contractors who include a major 'systems' component in their international construction activity, also seem to link back to their domestic economy in their materials procurement policies. Data for the 1979–83 period show that of the total contract value of their international work, between 20 per cent and 25 per cent was used for the procurement of building materials and components from Danish suppliers.[4]

5 Prospects for the Future

Given the fall in demand and changing geographical distribution of the international construction markets, the capacity of the Scandinavian contractors to expand or maintain their market shares must be evaluated. As a result of the changing structure of these markets, of the entrants from Newly Industrialized Countries, and the redefinition of their functions by many major international contractors, changes in strategies will be required for continuing success.

From the perspective of a contractor, new markets will always require a rethinking of strategy. Contracting is a high risk operation which requires a cautious approach to new markets. Even contracting for similar work in diverse locations aggravates problems of management. Changes in type of project may imply significantly different resource requirements, technical skills and managerial strategies. In international markets, changes are exacerbated by differing forms of contract, building regulations and standards, local construction practices, and providing a funding package for many projects. The Scandinavian international contractors have demonstrated their competence in coping with these problems; the issue, however, is their capacity to cope with the changes now underway in international construction.

With respect to the changing geographical distribution of projects, Scandinavians do not seem to have many problems. As was pointed out earlier, many Scandinavian contractors are already established in the presumably growing international construction markets of the US and South East Asia. Moves into the US are growing mainly through the acquisition of local contractors and initiatives in real-estate development. Many Scandinavian contractors have used both strategies to effect market entry, the most recent being the acquisition of E. W. Howell of New York by Selmer-Sande, Norway's largest contractor, to supplement the real estate developments in Florida of their subsidiary the Selmer Corporation. The Swedish contractors Skanska and ABV have adopted a similar bifurcated strategy to penetrate the US markets, developing a capacity for contract construction and real estate development in parallel.

Market penetration in South East Asia encounters few operational problems because of the relative strength of the local construction resource base, but it is hard because of the market leader role of the major Japanese contractors. The Scandinavian contractors have penetrated these markets as part of a deliberate long-term strategy. Christiani & Nielsen, for example, has been working in Thailand for many years and has also worked in Malaysia. ABV are shareholders in Balken Piling (Singapore) and Hume-Balken (Malaysia). These two cases, however, remain too small to qualify as a major penetration of a growing market.

The capacity of Scandinavian contractors to adjust to change is proven by both their domestic and international experiences. Danish contractors have been involved in consortia within Denmark and internationally, Swedish contractors have successfully operated in consortia for major international projects, constructed on a turnkey basis, and have participated in counter trading. Finnish international construction is founded on non-traditional methods of contracting. Collectively their international profile is characterized by flexibility and innovation in forms of contract and modes of trading. However, this experience does not guarantee success for the future; rather it provides a foundation upon which future developments can be based.

Despite the strengths of their domestic economies, each country cannot individually expect to match the potential for funding which exists in the US, Japan, the Federal German Republic, France and the United Kingdom. Equally they cannot expect to carry the same political 'clout' of these larger countries. This weakness might exclude their involvement on major projects in the future which are tied to funding. So far, in the field of major projects, the Scandinavian contractors have had a low profile. Where they have been involved it has typically been linked to a wider political strategy, such as the Finno-Soviet trade agreements, or with development assistance, such as Skanska's contract for the Kotmale hydroelectric power project in Sri Lanka. There is clearly only a limited scope for tying up markets in this way.

Their major international strengths have been in those areas of work where they have a demonstrable domestic competence; complex buildings of high quality, specialized engineering projects and technological innovations. The domestic base is fundamental in explaining past success and developing future strategies. Scandinavian international contractors are both the beneficiaries of well conceived programmes and strategies for domestic construction and major actors in the realization of these programmes.

There is a real sense in which each contractor separately is exporting a generalized commodity, 'Scandinavian construction', which is well received in the market place. It is not a coincidence that Scandinavian consultants have the third largest share of the international market, only exceeded by the US and the United Kingdom. Additionally the international success of Hifab and Ake Larsson, Swedish specialists in project and construction management, is also founded (in part) on their selling the product 'Scandinavian construction'.

6 Conclusions

Economic theory does not lend itself easily to the analysis of contract construction. Although some consider construction to be aggressively price competitive, the nature of the product and the complexity of the cost and production functions, both from the perspective of the client and contractor, makes it a very special case for analysis. Consequently it appears to be naive to look for any simple comparative advantage in explaining the international success of the Scandinavian contractors. More sensibly we should distinguish those factors which are unique to Scandinavian construction from those which are common to all international contractors, and examine the competitive advantages of the Scandinavians. Especially important are the special 'routes of entry' which eased them into the international construction markets.

Uniqueness is a rather amorphous attribute; we are all unique but are remarkably predictable in many given situations. To the extent that Scandinavian construction is unique it must relate to those factors which conditioned the development of the existing modern construction industry. Prior to 1950 the dominant theme of the Scandinavian construction industry was traditional; well proven technologies and skills were used to produce buildings and civil engineering in a predominantly labour-intensive way. This statement is not meant to imply that there were no sophisticated contractors, examples to the contrary have previously been given, but these contractors were the exceptions and were more involved in civil engineering than in building.

The expansion of their domestic economies and the social objectives of governments significantly increased demand for new construction in the Scandinavian countries, in a way that was recognized as beyond the capacity of the existing construction industry. As a deliberate act of policy new forms of construction were developed, new technologies introduced and the industry modernized. The industry which evolved from this process was characterized by concentration, vertical integration and a capacity, among the larger firms, to operate across a wide range of different projects and technologies, especially heavy prefabricated industrialized systems building. The Scandinavian international contractor has developed from these uniquely national circumstances.

International contractors are not homogeneous, and we cannot sensibly compare Skanska or Christiani & Nielsen with Bechtel or the Parsons Corporation. However we can compare them with Ballast Nedam and Wimpey. They have similar turnovers, ranges of projects and levels of international activity. The only evidence we have to claim any competitive advantages for the Scandinavian contractors is their significant presence in the international markets. We cannot argue that they are either more or less efficient than similar contractors from other countries. Scandinavians claim, with justification, a proven competence in construction management and technical sophistication. However for any given international project they still have to 'pare their margins' and take an increased risk to win the contract, the same as contractors from any other country.

Scandinavian contractors have certain special 'routes of entry' for international work. Finnish contractors for example have benefited from Finno-Soviet trading agreements. Construction in the Comecon countries has been assisted through proximity and the general political profile of the Scandinavian countries. However their competence and willingness to engage in counter trading has probably been more important. Another special 'route of entry' has been the degree of political and economic integration among the Scandinavian countries, which is of course assisted by the similarities of language in Denmark, Norway and Sweden. Consequently the level of intra-Scandinavian trade in construction was predictable.

An important 'route of entry' which is common to all the Scandinavian contractors is the development assistance policies of their national governments. These policies appear to be generous, directed to specific development objectives and specific countries. Many medium-sized Danish contractors have effected market entry through aid projects and Skanska's Kotmale project in Sri Lanka was funded by SIDA, the Swedish development assistance authority. Their influence is not only direct, these policies also build up a fund of goodwill which extends beyond the target countries. In an interview with a major Swedish international contractor I posed the question: 'What as a Swedish contractor makes you special to a client in a developing country?' The reply was interesting if unexpected: 'Swedish foreign policy.' Here I rest my case.

Notes

1 Frederick Forsyth, *The Fourth Protocol* (New York: Viking Penguin, 1984).
2 Information provided by the JCC Construction Group.
3 *Engineering News-Record*, 27 November 1986 (New York: McGraw-Hill).
4 Data supplied by the contractors associations in Sweden and Denmark.

9

Brazil

JOSMAR VERILLO

1 Introduction

The development of the Brazilian construction industry is associated
with the process of industrialization and state intervention in the
economy.

In the colonial period, 1500–1822, construction was a matter of
simple roads, docks, housing, shops, warehouses, churches, and
office buildings, a very crude process of moving earth and using stone,
wood, clay, bricks and slave labour. The state organized its own projects directly without general contractors, as did private clients.

The open door policy initiated in the year 1808, the coffee export
boom which started around 1850, the end of the slave trade in 1851,
the abolition of slavery in 1888, increased European immigration, and
the emergence of an infant textile and food processing industry all
brought important changes to the construction sector. Demand was
created for more housing, industrial installations and infrastructure
like railroads and ports. Formal construction firms appeared and
beginning in 1874 several engineering schools were set up.[1]

British firms predominantly took charge of building railroads and
ports. Brazilian firms were normally subcontractors but gradually
undertook more complex tasks. By the 1920s Brazilian firms were able
to build railways, but demand for that kind of infrastructure had
diminished.

In the 1930s the construction of dams to generate hydroelectricity
privately, created a new market for foreign firms. This market was,
however, very small until the 1950s when the hydroelectric sector was
nationalized. In the late 1940s Brazilian private construction enterprises remained short of capital, technology, and weak in organization. They prevailed in offices, housing, and road construction. The

large foreign firms concentrated their activity on the hydroelectric sector and other big contracts.

The Federal Government through the National Department of Roads (DNER) built most roads directly until late 1940s. In the 1950s the state withdrew from force-account road building, leaving it to the private sector, becoming at the same time the biggest individual buyer of construction services. Most of the firms currently exporting construction services were formed between 1950 and 1970 (Table 9.1). These decades were a period of rapid growth and consolidation that can be better understood in the light of the structural changes occurring in Brazilian society from 1930 until 1950 from which resulted the redefinition of the role of the state in the economy. After 20 years of growth and consolidation (1950–70), internationalization began in the 1970s. This rapid development can be attributed to the role of the state as a regulator, protecting the internal market for indigenous firms, and as a major client for construction work. In 1984 construction value added accounted for 5.7 per cent of the GDP and provided 3.1 million jobs for 7.2 per cent of the active working population. Indigenous capital prevails in the sector.

Overseas contracting is a very recent phenomenon in the Brazilian construction industry. Before 1973 there were just a few cases in neighbouring countries like Paraguay and Bolivia. The available ,

Table 9.1 *Sample of 54 Major Brazilian Contractors by Year of Founding*

Period	Number of firms
1921–25	2
1926–30	—
1931–35	1
1936–40	3
1941–45	3
1946–50	4
1951–55	10
1956–60	7
1961–65	11
1966–70	9
1971–75	3
1976–80	—
1981–85	1

Source: Brazil, *A Catalogue of Engineering and Construction*, Export Midia International Editora Ltda, São Paulo, 1986, major design and construction firms participating in export activities.

records from the Brazilian Association of Industrial Engineering show that the oldest construction contract abroad was made with Bolivia by Companhia Tecnica International (Techint) to supply a railroad bridge in 1960. Records of these transactions at the government level are practically non-existent (*Diagnóstico Nacional da Indústria da Construção*, vol. 16, page 35). The available data, worked out by the Brazilian Association of Industrial Engineering, ABEMI, shows 66 construction contracts before January 1984 (Tables 9.2 and 9.5). The majority (304) were design and feasibility studies. A remaining 71 referred to industrial installations and ancillary services.

Finding the value of contracts was impossible. Estimates from the Centre of Studies of Foreign Trade Fund (FUNCEX) put the total of design and construction services exported in 1980 at US $150 million.

Table 9.2   *Number of Contracts by Area According to beginning year of construction, Brazil*

Year	South America	Africa	Central America[a]	Middle East	Europe	North America	Oceania	Non specified	Total
Before 1970	11	1	4	—	—	—	—	—	16
1970	3	1	—	—	—	—	—	—	4
1971	2	1	2	—	1	—	—	—	6
1972	2	1	—	—	—	—	—	—	3
1973	9	1	1	—	—	—	—	2	13
1974	8	3	—	—	—	—	—	2	13
1975	10	11	2	—	—	—	1	3	27
1976	13	10	1	—	6	1	—	3	34
1977	14	2	2	1	2	—	—	3	24
1978	24	7	4	2	2	1	—	3	43
1979	21	9	5	3	1	3	—	2	44
1980	27	11	4	5	1	4	—	5	57
1981	21	8	2	1	—	—	—	1	33
1982	12	3	2	1	1	—	—	—	19
1983	7	4	—	1	2	—	—	2	16
1984	2	—	—	—	—	1	—	—	3
Not identified	48	25	5	7	—	3	—	1	89
Total	234	98	34	21	16	13	1	27	444

Source: Fundação João Pinheiro, Diretoria de Projetos I, *Diagnóstico Nacional da Indústria da Construção*, Belo Horizonte, 1984, vol. 13, p. 20, and ABEMI, Annual Report, 1983.

Note: [a] Includes Mexico.

Goods exported as a result of the contracts may add an additional $500 million. A gross estimate of the value of services exported in previous years was worked out in a study conducted by FUNCEX. The numbers are not impressive, below $120 million in the Central Bank account and an average of $151 million in the ABEMI estimation for the years 1978–80 (in 1982 dollars). The Central Bank records only actual currency repatriated. It must be taken into account, however, that the major Brazilian contractors opened foreign subsidiaries partially in order to avoid bringing currency earned on foreign contracts into the country. Of 59 firms exporting services (design and build) which advertised in the *Brazilian Catalogue of Engineering and Construction*, 14 mentioned foreign branches. Once they repatriate these earnings, firms must obey official priorities and rules for spending hard currency, which greatly reduces flexibility.[2]

There is no financing in hard currency by the Brazilian government or private organizations, and it takes time for approval of any expenditure in hard currency. If a firm really wants to compete for contracts abroad, it does not want to be held back by such restrictions. Another disincentive to repatriating foreign currency is that the official exchange rate is usually overvalued more than 30 per cent. For all these reasons external accounts fail to record a good share of the transactions made by the foreign subsidiaries of Brazilian firms, especially on services.

The *Engineering News-Record* (*ENR*) annual survey of the 250 top international contractors (Table 9.3) provides some insight about the behaviour of the major Brazilian contractors. Nine contractors appear off and on among the top 250 firms. Construtora Mendes Junior was most consistently among the two or three Brazilian companies listed in any one year. With six or seven firms below the *ENR* cut-off, those included make up a small and erratic but perhaps not useless sample. After a peak of US $2.2 billion in 1981 in new contracts, there was a decline of 98 per cent to US $34 million in 1986.

The Brazilian government through two funds, Carteira de Comercio Exterior do Banco do Brasil (CACEX) and Fundo para Financiamento de Estudos e Projectos (FINEP), finances design, feasibility studies, materials, and equipment for foreign contracts. The records of these operations are available and provide another way of measuring the level of activity of the Brazilian construction industry abroad. The outstanding amount of loan contracts in 1985 was $1.9 billion, and about $700 million for feasibility and preliminary studies. The amount financed is normally between 30 and 80 per cent of the value of the project and goes only for goods and services purchased in Brazil. The amount contracted is well above the amount of outstanding loans. CACEX and FINEP lend Brazilian currency to the contractors and receive repayment in hard foreign currency. It is estimated that in the period

Table 9.3   *Brazilian Firms Among the Top 250 Contractors
of ENR, 1980–85
(US$ million, 1982 values)*

Year	Contractor	US$ total	US$ foreign	Percentage foreign	Countries
1980	Constr. Mendes Jr.	3,362	1,726	51.3	Colombia, Uruguay, Iraq
	Tenenge	261	55	21.3	Chile, Paraquay
	Norberto Odebrecht	278	35	12.5	Chile, Peru
1981	Constr. Mendes Jr.	4,130	1,993	48.2	Colombia, Paraguay, Ecuador, Iraq
	Norberto Odebrecht	1,577	201	12.7	Bolivia, Peru
1982	Paranapanema	—	588	—	Nigeria
	Constr. Mendes Jr.	322	198	61.4	Colombia, Paraguay, Uruguay, Iraq, Sub-Sahara
	ECISA	190	155	81.5	Paraguay, Tanzania
1983	Norberto Odebrecht	770	577	75.0	Peru, Angola
	Andrade Gutierrez	432	226	52.3	Bolivia, Ecuador
	Constr. Mendes Jr.	81	53	64.7	Colombia, Paraguay, Uruguay, Iraq, Sub-Sahara
1984	Constr. Mendes Jr.	926	328	35.4	Colombia, Paraguay, Iraq, Sub-Sahara
	Montreal Engenha	251	81	32.3	Greater Antilles
	Guaranta	—	37	—	Algeria
1985	Andrade Gutierrez	439	139	31.6	Bolivia, Ecuador, Congo
	Affonseca	—	16	—	Chile

Source: *Engineering News-Record*. The Top 250 International Contractors, July 1981 to July 1986.

1982–4 Brazilian firms signed contracts worth US $3 billion (*Brasil Comércio e Indústria*, July–September 1985, p. 40). The major Brazilian contractors had 30 per cent of their earnings from contracts abroad while smaller firms earn only about 6 per cent of their earnings abroad (FUNCEX, *Relatório Básico*, part 1, p. 85).

There is no measure of employment created at home as a result of foreign contracts. A rough estimate puts between 2.0 and 4.5 per cent of the workforce employed by the heavy construction industry as working abroad in the period 1981–4 (Table 9.4). There is a consensus among entrepreneurs about taking only a necessary minimum number of unskilled and semi-skilled Brazilian workers abroad. They argue that the Brazilian worker ends up being more expensive than host

Table 9.4 *Brazil, Heavy Construction Industry,*
Number of Employees, 1981–6

Levels of formal education

Year	Superior	%	Inter-mediate	%	Others	%	Total	Abroad	%
1981	13,885	5.1	35,286	12.7	227,710	82.2	276,881	5,900	2.13
1982	13,694	5.6	31,152	12.6	201,688	81.8	246,534	5,937	2.40
1983	12,219	6.0	23,793	11.7	167,632	82.3	203,644	5,937	2.90
1984	13,261	6.5	26,071	12.7	166,056	80.8	205,388	8,903	4.30
1985	n.a.	n.a.	n.a.	n.a.	n.a.	n.a.	251,017[a]	n.a.	n.a.
1986	n.a.	n.a.	n.a.	n.a.	n.a.	n.a.	265,336[b]	n.a.	n.a.

Source: ABEMI, Annual Report 1985, and Field Research.
Notes: [a] Estimated from the *Brazilian Catalogue of Engineering and Construction*, 1986, Export Midia International Editora Ltda, São Paulo.
[b] Estimated from *Dirigente Construtor*, July 1986, Vol. xxii, Editora Visao Ltda, São Paulo.

cost of taking an unskilled Brazilian worker to Iraq is $3,000, for example. Except where there are no local workers available and no other cheap source of labour, the only Brazilians at the work site are administrators and key technical personnel. In some cases Brazilian firms hire Chinese, Turkish, Portuguese and other nationalities. In Iraq, for example, Construtora Mendes Junior signed a contract for labour from China.

The most important markets (by number, not value) for Brazilian firms until 1984 were Latin America with 64 per cent of contracts; Africa with 23 per cent; and the Middle East was third with 5 per cent (Table 9.5). The Middle East contracts had a much higher value. The Construtora Mendes Junior alone contracted with Iraq services in the amount of US $1,992.3 million in the period 1979–84.

The country's legal and fiscal institutions are still not well prepared to handle foreign contracts. Despite a pledge to support exporting firms, the government has not developed a medium or long run policy. The natural advantage for exporting construction services could be enhanced through improving the legal and fiscal framework, the eliminating of excessive paper-work, increasing financial support to smaller firms, and developing transportation facilities to new markets in Africa and Latin America.

There are now few foreign enterprises operating in Brazil (Annibal Villela, 1981a). Market protection still prevails. Foreign contractors are allowed to operate only in fields where there is no local capability. All projects must have authorization of the National Institute of Industrial Property (INPI), an entity linked to the Ministry of Industry

Table 9.5   *Foreign Contracts by Type of Service
and Geographic Area
(before January, 1984)*

Geographic areas	Type of service				
	Design and feasibility studies	*Heavy civil construction*	*Industrial installations*	*Ancillary services*	*Total*
South America	154	38	24	18	234
Africa	70	16	3	9	98
Central America	23	6	3	2	34
Middle East	13	6	—	2	21
Europe	13	—	3	—	16
North America	12	—	—	1	13
Oceania	1	—	—	—	1
Non-identified	18	—	2	4	24
TOTAL	304	66	35	36	444[a]

Source: Fundação João Pinheiro, Diretoria de Projetos I, *Diagnóstico Nacional da Indústria da Construção*, Belo Horizonte, 1984, vol. 13, p. 13.

Note: [a] Three contracts had no specification of country and type.

and Commerce. Each project is analysed together with a constructors association like ABEMI. Only when the local constructors say they cannot do the project is authorization given to a foreign contractor. Even then association with a local firm is normally required (*Diagnóstico Nacional da Indústria da Construção*, 1984, vol. 2, p. 53). Exceptions are projects financed by international organizations where the contract is decided by open bid. Many foreign contractors, however, have learned how to become a 'national' and are operating in the market at different levels of partnership and association with local firms.

2   Business Strategy: Contracts, Co-operation and Finance

The majority of Brazilian firms working abroad were founded between 1950 and 1970, and they are today incorporated with stock traded on the stock exchange. The five major exporting constructors initiated activity between 1945 and 1955. Their first major contracts abroad were signed between 1969 and 1984. On the average they took three to five years to get the first contract after the decision to go abroad. Four of the nine appearing in *ENR* top 250 have foreign subsidiaries.

In the 1950s changes in Brazilian society led to: (1) a strong option for industrialization; (2) the adoption of planning with targets, the 'Programme of Goals' of President Kubistchek; (3) withdrawal of the state from building activity and financial autonomy for the DNER (National Department of Roadways); (4) creation of Petroleos Brasileiros s/a (PETROBRAS) and state monopoly in the oil industry; and (5) National Plan for Electrification. In short, state intervention in the economy was aimed at all the sectors where private capital was lacking. Many of today's big contractors were formed or consolidated in this period. It is interesting to notice that most of these firms were formed by the association of two or more engineers that wanted to start their own business. Important for these contractors was the road building boom initiated in 1945 and lasting until 1955. The growth of these companies from that time on was the result of the massive investment by the Brazilian government in infrastructure work, hydroelectric power stations, and oil exploration facilities. When this market showed signs of exhaustion, they started to look for contracts abroad. Some enterprises had the idea of building abroad well before the Brazilian government cut investment as a result of the oil crisis. When these cuts came they were ripe for taking chances abroad.

Currently Brazilian contractors are looking for any type of work. They are determined to have a strong foothold outside the country and to be capable of quick mobilization of expertise. For example Mendes Junior won a contract to build a 1,100 kilometres railroad in Iraq without having recent significant experience in this kind of work. In order to do the job they hired the majority of the technicians of Volta Redonda steel mill who had the know-how and were experiencing idle capacity. Norberto Odebrecht is working with the Russians in Angola. To give assistance to the Angolans to oversee the construction of the Capanda Hydroelectric Plant, a Brazilian State enterprise (Centrais Electricas de Furnas) with no experience abroad was hired. In Venezuela, Camargo Correa worked in partnership with the American contractor Guy F. Atkinson. So expansion can be observed on a variety of fronts.

Brazilian firms prefer to work in Latin America, Africa and the Middle East. Some of them are approaching the Chinese market. They argue that developing countries are their natural market. Their more appropriate technology and similar aspirations give them an advantage compared with contractors from developed countries.

In spite of winning several open bid projects financed by the World Bank and the Interamerican Bank for Development, Brazilian contractors prefer direct negotiation. In this strategy they can offer Brazilian government financing as a big advantage. That is one of the reasons why government connections are so crucial to their strategy. Also they can help give the inside track when a project is announced.

Contractors see favourable conditions in Africa where many

countries are anxious to distance themselves from the former European colonial rulers. At the same time the Africans do not want to improve relations with the United States because of its past policy towards their countries and apparent support of South Africa. In Congo-Brazzaville for example, the American Ambassador perceiving the easy relationship experienced by the Brazilians, explored the possibility of Andrade Gutierrez associating with American contractors in other projects. This should come as no surprise. Brazil, being a Third-World country relatively distant from the East–West conflict, with development aspirations and something to offer, makes the diplomatic effort easier (Annibal Villela, 1981b). People in the Third World are growing increasingly weary of being caught in the East–West conflict. According to the entrepreneurs, the current Brazilian foreign policy is favourable to their business, and the Brazilian Ministry of Foreign Relations (Itamaraty) is doing an excellent job of supporting contractors abroad. Some entrepreneurs admit that they sometimes go beyond the initiative of the Itamaraty. They have engaged successfully, with the participation of government officials, in complex barter deals with Nigeria and Angola, among others. This kind of transaction necessarily involves strong government participation. The product normally received by Brazil is oil, whose exploitation and refining are a state monopoly.

In the direct negotiating strategy firms must anticipate the needs of the host country and present a package including design, construction, and financing. For them the most efficient manner of getting the necessary lead time is to maintain a permanent presence in prospective markets because information systems through official channels are ineffective. When a project is publicized most of the arrangements have already been made. This was learned after trying to participate several times on short notice in open bids already designed to favour a given contractor. The contractors are trying to work close to the governments of host countries to know first hand what they are planning, to help them to arrange financing, to assist them in the pre-design stage of the project and so on. All the major Brazilian contractors keep permanent offices in the countries where they have projects and in countries which represent potential markets. This does, however, increase the pre-contracting costs. They also see this as essential for gathering information in the preparation of the bid. A close contact can help to get direct negotiations instead of open bidding. The strategy of direct negotiations which normally involves counter trade works better with those countries with which Brazil has a trade deficit or political interests. The entrepreneurs see an increasing government role in the process of contract awarding, even in open bid contracts. Governments, according to them, can put pressure on international financial organizations.

Brazilian contractors compete successfully internationally in areas where the technology is standard, and there is a wide margin for capital/labour substitution. These areas include road construction,[3] railroad construction, dams, sewerage systems, office buildings, bridges, airports, housing projects, some industrial installations, and power lines.

One interesting contract was being negotiated in 1986 between a Middle Eastern country and a Brazilian international consortium. The host country will pay the contractor for local expenditure in local currency, but the portion of expenditure in hiring foreign workers and buying foreign products will be paid in cruzados. Only the administrative fee (profits and overhead) will be paid in dollars. An exchange rate between the country's currency and the cruzado was fixed during the period with allowance for inflation. In order to pay the contractor in cruzados, the Middle East country could accept cruzados as payment for oil sales. This would seem to limit the flexibility of hiring workers. Brazilian low-skill workers being too expensive, it is likely the consortium will seek another Chinese-type labour contract with payment made in cruzados. This kind of deal must involve a guarantee against inflation, but it leaves hard currency, always a shortage in developing countries, out of the transaction.

Brazilian contractors also have had some bad experiences in winning bids for construction against the will of the designers of the project. They have learned that a designer acting as consultant to oversee the construction process can put the constructor out of business if they wish to. This may explain why they have showed a preference for turnkey and design-build type of contracts. The major firms in Brazil are designers and builders. When they get a contract and their internal capacity is not enough to design the project, they subcontract the job.

Three of the major contractors have evolved into conglomerates with subsidiaries in mining, oil drilling, industrial installations, steel mining, machine tool manufacturing, trading, air transportation, chemicals manufacturing, and other activities. These three are Norberto Odebrecht, Mendes Junior, and Andrade Gutierrez. Most contractors have had associations with host country firms and third country firms but rarely with other Brazilian firms. Some experts argue that the competition among Brazilian firms for foreign contracts is predatory, and duplicates effort. Initiatives to create consortia of builders to present only one good proposal, have failed.

One interesting arrangement regarding food in a contract in Iraq is worth mentioning. During the construction of a railway in that country the contractor Mendes Junior had to prepare 39,000 meals a day for 50 different nationalities of workers. They had their access to hard currency limited because of the Iran–Iraq war, so the capacity to buy in the international market was impaired, and Brazilian suppliers were not able to supply the required variety of foods either. The problem was solved by what they called 'triangulation'. European trading companies supplied the required food items and the payment was made by selected Brazilian products. That is the way the contractor entered the trading business. Before that the job of the trading company in the group was to facilitate equipment imports.

3 Project Execution: Workers, Subcontractors, Materials, Equipment and Technology

The major contractors have similar human resource policies. The decision system is relatively decentralized, around the important figure of the contract manager. He has great power to conduct the work in his project.

In their first contracts abroad Brazilian firms normally took to the work place a high percentage of nationals, including low-skill workers. In some countries like Mauritania and Iraq, for example, there were no local workers available in the required numbers and the use of Brazilians was essential. In South American countries this problem does not occur. They normally have a trained or easily trainable workforce. In the construction of the hydroelectric power plant Charcani V in Peru the contractor is using only 9 Brazilians in strategic positions (*O Empreiteiro*, volume XXII, November 1985, São Paulo, p. 60).

This policy of transporting Brazilian workers seems to have changed. The initial plan of Mendes Junior in Iraq predicted the transportation of more than 12,000 Brazilian workers at the peak of the construction work, and the budget included the purchase of three large aeroplanes. But the number of Brazilian workers reached only about 6,000 because the cost of transporting Brazilian workers turned out to be higher than expected. Some entrepreneurs argue also that Brazilian workers do not adapt very well outside the country. The current policy seems to be to hire and train most of the workforce locally. Another option is to look for labour contracts from which no liability results to the contractor. In the Chinese-type contract any worker's complaint is directed to the Chinese government, not the contractor. That is not the case with directly hired workers. Some complaints have been filed in the Brazilian Labour Courts, increasing the costs of taking workers abroad.

Brazilian workers abroad are normally paid a maintenance share of their total salaries in the currency of the host country, and the rest is deposited in personal bank accounts in Brazil. Only the top level staff have free disposal of their salaries, as they receive the full amount in hard currency. This is one factor lowering the cost of unskilled labour, which, on the other hand, makes low-skill workers not eager to work abroad. They have their purchasing power tightly limited abroad and once they get home their remaining family may have spent salaries deposited in the bank. That may not be an attractive situation unless the local job market is in a slump.

For the enterprises, when the project is financed by Brazilian institutions, they can only pay workers in cruzados. Many contracts would not exist if not financed by Brazilian entities. We must bear in mind that when the Brazilian economy heads towards full employment the strategy of giving loans in cruzados and exporting local skilled labour becomes difficult to sustain, because priority is given to internal projects. The first signs of that could already be seen in 1986. In the words of one entrepreneur: 'We are so busy with local projects that we are not interested in going through the trouble of seeking work abroad.'

Entrepreneurs in general assert that they place no restrictions on the formation of trade unions, but that statement is not easy to verify with the employees. It is well known that in this kind of activity it is difficult for labour to organize. Workplaces are very far apart and communication between members is difficult. Organization by workplace allows easy identification of the activist members and they can be dismissed or transferred to other places where they do not present a danger of disrupting activities.

Trade unions in Brazil in general are not strong, as they were suppressed by the military government that lasted from 1964 until 1985. So contractors have no experience in negotiation and labour disputes. Camargo Correa, a major Brazilian contractor had labour problems on the Guri Dam in Venezuela where unions are stronger as democratic regimes have prevailed since the 1950s. After a series of problems the US firm, Guy F. Atkinson, took over and Camargo Correa continued as a subcontractor.

Contractors acknowledge no specific provisions or open pressure for the use of local labour or a more appropriate technology by host countries. Most countries, however, have laws specifying the share of foreign workers allowed to work in the same organization. It is the opinion of some entrepreneurs that if this pressure was necessary the contract probably would not be awarded in the first place. This is a subtle question that the contractor has to anticipate. In order to present the proposal the contractor has to gather a great deal of information and he knows what the host government is expecting the terms of labour use and labour training, so he puts it into the proposal. They also increase their chances of being awarded the contract if they associate with local firms. Brazilian firms have normally associated with local firms whenever they can find a suitable partner. These associations have not been particularly painful because of the cultural kinship of most countries where Brazilian contractors are operating (Latin America and Africa).

Most standardized equipment used in construction work can be purchased in Brazil. The share of imports of the sector in total Brazilian imports is very low when compared with other sectors of the economy. But most contractors do not have to make the decision to use Brazilian products or imports. Since the financing comes normally in cruzados the only choice is buying Brazilian products. Even when the financial restriction is not binding, there is a 'Law of Similars' forbidding imports of equipment for which local production exists or can be easily started. The only way of getting round these restrictions is to channel the contract to a foreign subsidiary if it specifies payment in hard currency and no Brazilian financing is involved.

Firms exporting design and construction services are financially the industry's strongest. This gives them a strong bargaining position and purchasing power in the Brazilian market. That means they get the most attention from the suppliers of building materials. There seems to be no significant market failure in this area. The contractors are financially capable of internalizing any sector of the building materials industry they wish, but their diversification did not necessarily take place in this area.

4 Government Policy

The Brazilian government has no national policy to promote the export of design and construction services. Construction projects of the turnkey or design-build types benefit from existing manufactured goods export promotion incentives. No significant additional mechanisms have been introduced. Support is given to exporting firms on a case-by-case basis, which requires much innovation because many situations are not contemplated in the laws and rules of the country.

The clients for heavy construction projects are normally governments, which makes bilateral relations a very important factor in the contractor's market expansion. Entrepreneurs agree that the Brazilian government has been doing a very good job in this area. The big enterprises get what they need most of the time. The medium and small contracting firms do not always get the support they need because of institutional inadequacies, and they are crowded out in the dispute for attention from lending institutions and the Itamaraty (Ministry of Foreign Relations).

Brazilian firms get a number of incentives to export services:

1. All purchases of Brazilian goods and equipment shipped abroad for construction projects are treated as a normal export operation. They are exempted from the payment of the industrialized products tax (IPI) and the merchandise commercialization tax (ICM). If equipment returns to the country at the end of the project, it will be treated as an import and tax will be assessed over the residual book value. Imported goods and equipment without a similar national product can be excluded from the payment of taxes as well if they are going to be used in construction abroad.

2. Firms selling engineering, architectural, designing, consulting, and ancillary services abroad can deduct from the net total profit a percentage of the profit earned abroad for corporate tax assessment. The percentage is determined by exported services as a share of total sales. The amount deducted from the profit cannot exceed the amount of hard currency actually brought into the country.

Earnings from technical services, technical assistance, administration fees to foreign consumers paid in hard currency can be deducted from the operational profit for the assessment of corporate taxes.

Brazilian workers living abroad, employed by subsidiaries, agencies, or branches of firms at least 5 per cent owned by Brazilian nationals, can choose to pay taxes in the host country. In this case his income will not be taxed in Brazil.

At the municipal level there is no exemption of taxes on exports. The most known tax affecting firms at this level is the tax on services (ISS).

In December, 1982 a new law regulating the situation of the Brazilian worker abroad was passed. Basically the law establishes that: a) the salary must be stipulated in cruzados, independent of the currency it is going to be paid in. Conversion and remittance of the salaries at the (usually overvalued) official exchange rate are allowed while the worker is abroad; b) the period worked abroad is counted for social security purposes in Brazil; c) the contractor is allowed to deduct from the salaries and other dues to employees in Brazil any severance payment made to comply with host country laws; d) during the period the worker is abroad the contractor is exempted of some social security contributions like Education-salary, Industry Social Service, Commerce Social Service, National Service for Commercial Training, National Service for Industrial Training, National Institute for Colonization and Land Reform; and e) Brazilians in order to be hired by foreign firms must ask for permission of the Ministry of Labour. This permission is given only for working in firms owned at least 5 per cent by Brazilians.

3. Export financing:

(a) Preliminary financing: for services exports. It is meant to finance the purchase of equipment and services in the country to fulfil contracts abroad. It is made by CACEX.

(b) Financing of trade promotion: it consists of loans for study programmes, research of external markets, research of techniques of commercialization, promotion and advertisement of national products, participation in fairs, and elaboration of proposals or projects for international bidding. The interest rate charged is normally below market interest rates.

(c) Financing of exports of goods and services. This is meant to give the firm or the client the option of extended payment in large projects. It can be made in two versions, by CACEX: first, supplier's credit, with a time limit of over six months, through documentation of the sale of goods and services; or second, buyer's credit made through the discount of the promise of payment made by the customer.

(d) Advances, either partial or total, on future exchange transactions of exporting firms in cruzados or allowed hard currency expenditure based on contracts signed by clients. This kind of thing originates only from hard currency control. Sometimes the contractor is awarded the contract but the instalments are due at some time in the future. So the client makes a future exchange contract with the Brazilian bank in order to allow the contractor to receive an advance of foreign currency or even local currency.

(e) Interest rate equalization system. This is intended to give incentives to private banks to finance exports. The long-run interest rate used by the Banco do Brasil to finance projects is about 7 per cent. If private banks contract loans abroad at interest rates above that and use this resource to finance exports at the 7 per cent interest rate, FINEX will pay the difference between the interest rate paid to international lenders and the 7 per cent plus an administrative fee. This applies to loans for financial studies, technical assistance projects, expenditure abroad in order to win contracts (limit of 10 per cent of actual hard currency brought into the country as a result of the contract), and insurance to cover broken contracts or lack of payment by the customer. This kind of financing, however, is not very effective. Most firms do not know about the programme. Those who know it argue that there is too much paper work involved and the programme becomes too uncertain. For some types of expenditures the contractors are demanding a form of grant.

(f) Other entities participating in the financial support of design and construction services exports:

- Secretariat for International Technical and Economical Co-operation (SUBIN): Programme of Technical Co-operation With Developing Countries: It finances elaboration or execution of projects or activities of Brazilian technical assistance and co-operation with other developing countries.
- Studies and Projects Fund (FINEP): Programme of Support to National Consulting Firms: Finances research for markets, proposal making, and feasibility studies (mentioned above).
- Brazilian Centre for Support to Medium and Small Enterprises: Programme of Support for Consortium Formation: It pays up to 80 per cent of the necessary operational expenses to install and maintain consortia of small firms to export goods and services.

Apart from fiscal exemption and financial support, Brazilian contractors get support from the Brazilian Ministry of Foreign Relations. Besides the support given by the diplomatic service in the form of advice, visits to clients, and participation in negotiations, there is a system of information established. The 'Projects Information Bulletin' is meant to give preliminary information about projects that will take place in several countries. The 'Summary by Sector' and 'Technical Description of Projects' give details of projects applying for financing to international organizations, an historical account by countries, which give the entrepreneur both a general picture of the sector and details to elaborate the bidding proposal. In many urgent situations, however, a telex system is used, and contractors get information the same day.

The participation of diplomats in negotiations and visits involves a change of mentality that is still taking place at the Ministry of Foreign Relations. Diplomats normally want to deal with politics. They still see business as the dirty part of their job. The change of approach may take several generations because diplomats begin their career early in life in a hierarchical organization and their training is not business-oriented or strong in its economic aspects.

In spite of it not being properly institutionalized, we can see that there is a set of organizations and measures aimed at the support of construction exporting firms. Entrepreneurs however demand more financial support, complete exemption of corporate taxes on profits earned abroad, financing in hard currency, allowing employees to bring into the country duty-free the goods bought while working abroad, inflation insurance, exchange rate insurance, and exemption from payroll taxes. The Brazilian contractors also complain that the Brazilian government is not putting enough pressure on international lending organizations to award contracts to Brazilian firms in the same proportion as the country's contribution to those organizations.

As for insurance, a system of export insurance was created in 1965. In 1967 insurance in hard currency was introduced. The agency responsible for that is the Brazilian Reinsurance Institute (IRB). This insurance covers contract obligations and credit. The credit insurance covers lack of payment for bankruptcy, political factors and others. Contractors claim however that insurance premiums are too high and many firms do not make use of it.

It is the responsibility of Banco do Brasil, or authorized banks to give the required coverage to Brazilian contractors to operate abroad. This includes bid bond, surety bond, performance bond, and refunding bond. No charges are made for the concession of these guarantees. Collateral is required, and normally it consists of deposits, duplicates, promissory notes, export credit insurance, sureties and mortgages. Here the big contractors have a sweeping advantage. With their bargaining power and influence, anything is accepted as collateral. That is not the case with the small contractors and design firms that have few physical assets.

Entrepreneurs of the construction sector are unanimous in pointing out the protection of the home market for Brazilian firms as the most important factor for the development of the industry. Some of them have acquired a capacity to export services, but they still want to keep the local market protected. That is more important than competing abroad. They argue that foreign firms have technological advantages, and they would never transfer technology if regulations were abandoned. At the same time, from the evidence we have, the major Brazilian contractors have preferred in the past to diversify activities instead of investing heavily in the development of construction technology.

The government has no control over contractors exporting services. It does not oversee quality or the behaviour of contractors abroad. That may be one reason behind the difficulty in getting information. This lack of control or co-ordination may also result in excess of competition among Brazilian firms.

5 Concluding Remarks

The major Brazilian contractors are generally doing well. But there is a big gap between them and medium/small contractors and designing firms. The lack of institutional mechanisms harms small contractors. They cannot afford the pre-contracting expenses like travel abroad, market research, preparation of a competitive bidding proposal, maintenance of offices in prospective markets, and paying salaries to qualified personnel. Financing for this stage of foreign contracting is precarious in Brazil. With the exceptions of projects financed by international lending organizations, the nationality of the builder and equipment normally follow the nationality of the designer in Third-World countries unless firms of those countries are protected, as in Brazil. Pushing designers abroad in order to increase the probability of bringing contracts home requires large resources. Designing firms have no significant physical assets as collateral to guarantee financing. They would like grants for lost proposals, but according to government officials this is not feasible. Even the existing lines of credit are inflationary as they increase the state budget which in turn is heavily financed by printing money. So, in order to increase the volume of contracts Brazilians are relying heavily on politics. The participation of the state does not stop at the support of private entrepreneurs abroad. They are pushing the state enterprises like Petrobras and Interbras which have already signed several contracts abroad. The role of the state is also important in barter deals which do not involve financing. This mechanism is a very important way of making up for the lack of finance.

Another problem affecting the development of the Brazilian construction industry is the relative isolation of the country. Transportation is precarious between the country and its main markets, Latin America and Africa. It is cheaper to transport something from the United States or Europe to them than from Brazil. There are no railways, roads, shipping lines, or accessible air transportation. This increases the costs of Brazilian builders and it is an area requiring quick government action according to the contractors.

If we are to characterize the development of the Brazilian construction industry we would say that before 1950 there was an incipient large scale industry. The development stage went from 1950 until 1970. In 1970 the internationalization process accelerated and has been going on ever since. This seems to be the most difficult stage. The state led the firms during the development stage as a regulator and major client. Now at the internationalization stage the government is being pulled into institutionalization by contractors. The government still does not have statistics for the export sector. In spite of the billions of dollars signed in contracts in recent years, the services item of the external accounts has barely changed. The economic situation of the country and the restrictions related to spending of foreign currency discourage repatriation of funds. So we do not know to what extent the country actually benefits from exports of design and construction services.

Notes

1   Escola Politécnica do Rio de Janeiro (1874), Engenharia de Ouro Preto (1876), Escola Politécnica de São Paulo (1894), Engenharia Mackenzie (1896).
2   See Informação Semanal Cacex, Banco do Brasil, year 19, no. 921, Rio de Janeiro, 5 November 1984, p. 10.
3   In a recent open bid for a road construction in Ecuador, from the list of 77 bidding firms 9 were Brazilian. See Informação Semanal CACEX, year 19, no. 921, Rio de Janeiro, 5 November 1984, p. 10.

TUNISIA 4420
4210
6340

10

Tunisia

RIDHA FERCHIOU

There are about 1,300 construction firms in Tunisia, but only 53 of these are large enough to undertake works worth a million Tunisian dinars (US $830,000 1987 values) or more (see Table 10.1). Foreign contractors, therefore, participate in major projects such as road construction, hydraulic engineering, harbours, and airport construction. On the other hand, some large Tunisian contractors, having 20 to 30 years of experience, have extended their activities to the two neighbouring countries of Algeria and Libya. In these two countries they compete with large international contractors, mainly in low-cost housing construction. We shall consider in turn:

(1) the relative importance of exports and imports of building services;
(2) the experience of foreign contractors in Tunisia;
(3) the experience of Tunisian contractors abroad.

1 The Relative Importance of Exports and Imports of Construction Services in Tunisia

It is not easy to know exactly what is the contribution of foreign contractors in the total construction activity of the country unless we study all projects, one by one. Official statistics do not distinguish between international contractors and Tunisians. Projects are classified according to sector, the amount of investment, the source of finance, the employment created, but never according to the nationality of the contractor. However, in the Tunisian Balance of Payments under services exports and imports, there is an item called 'major works' (grands travaux) which takes into account all the current receipts or

Table 10.1   *Tunisian Construction Firms, June 1986,*
Annual Capacity

Regional distribution	Equal or less than 100,000 dinars	Between 100,000 dinars and 1 million dinars	Equal to and more than 1 million dinars	Total
Tunis and the Northeast regions	431	91	37	559
Northwestern regions	176	12	3	191
Rest of the country	535	54	13	602
TOTAL	1,142	157	53	1,352

Note: Tunis area includes 40 per cent of the total number of all construction firms and 70 per cent of large ones; in 1987, the value of the dinar was US $0.83.

current expenditures with the rest of the world related to construction activities (see Table 10.2).

From Table 10.2, we notice that 'major works' imports increased from 53 million dinars in 1981 to 69.6 million dinars in 1985 and represent about 15 per cent of total services imports in 1981 and 11.6 per cent in 1985. Major works imports are relatively steady, between 50 and 70 million dinars (1.00 dinar = US $0.83) These values are important because they give an idea about payments to international contractors that were not spent in Tunisia.

In general, international contractors remit from Tunisia to their home countries their profit, their overhead expenses, the amortization of their equipment, and a part of the wages and salaries given to skilled non-resident workers. They only spend locally what they have to spend: wages for local labour, local subcontracting and purchases, local taxes and insurance. It is not easy to say how much foreign contractors spend in Tunisia and how much they export to their home countries. It depends on the nature of the project, on the amount of local subcontracting, and on the terms of the agreement.

From Table 10.2, we notice that 'major works' exports increased from 7 million dinars in 1981 to 19.4 million dinars in 1985, with a peak of 34.8 million dinars in 1984. In relative terms, they only represent 1.4 per cent of the total services exports in 1981 and represent 2.3 per cent in 1985, with a peak of 4.6 per cent in 1984. We also notice that Tunisian exports of construction services were relatively low until 1983 and then rose fast. The average rate of increase during the whole 1981–5 period was 29 per cent a year.

Table 10.2   *Tunisian Balance of Payments*
Current Account Balance, 1981–5
(Millions of Dinars)

	1981	1982	1983	1984	1985
I Current receipts					
Merchandise exports	1,207.9	1,166.5	1,255.7	1,379.8	1,481.8
Services	516.5	611.9	703.5	749.4	829.5
(Major works included)	(7.0)	(8.0)	(18.3)	(34.8)	(19.4)
Incomes	219.9	273.1	285.7	287.8	242.4
Current transfers	18.0	22.1	19.9	17.4	17.2
TOTAL	1,962.3	2,073.6	2,264.8	2,434.4	2,511.9
II Current expenditures					
Merchandise imports	1,686.4	1,852.7	1,984.6	2,275.4	2,141.1
Services	356.2	448.4	499.4	539.8	600.9
(Major works included)	(53.0)	(70.0)	(64.6)	(51.2)	(69.6)
Incomes	151.0	172.3	183.4	231.1	273.1
Current transfers	9.0	8.3	11.8	8.9	8.3
TOTAL	2,202.6	2,481.7	2,679.2	3,055.2	3,023.4

Source: Tunisian Central Bank, Annual reports.

Between 1981 and 1985, the average deficit in terms of construction
services exports and imports was about 45 million dinars per year, but
in relative terms, this deficit is decreasing.

2   The Experience of Foreign Contractors in Tunisia

The participation of an international contractor in Tunisian projects
is possible whenever it is the case of an open adjudication or bidding.
This type of bidding is open to national and foreign contractors alike.
On the other hand, restricted adjudications are limited to Tunisian
resident contractors. However, some of these are affiliated to large international contractors (mainly French). Most of the adjudications are
restricted (80 per cent), however, the most important ones are open
to international contractors.

Projects Where International Contractors Can Participate
In general, public biddings or adjudications are open:

(1)   when the project is partly or totally financed through bilateral or multilateral aid programmes by international sources, mainly the World Bank. Open adjudications are required because they are supposed to give more competition and consequently better prices.
(2)   when there is no specific source of finance for the project, and the state or the client expects the interested contractor to offer a source of finance. This is generally a typical case of tied aid: the international contractor provides the client with credit from one or more banks of his home country, which the Tunisian government guarantees.
(3)   when the project is novel to Tunisia and requires some technical skills that the local Tunisian contractors do not have. This case of technology transfer appears whenever the project is very large: a big dam, port construction, power station building, etc. . . .

The three cases often occur at the same time. A large project is very often new in the country and requires know-how. It is generally expensive and requires domestic and international sources of finance.

Whatever the project, the selection of the contractor (national or international) is done on the basis of:

(1)   the importance of the firm in terms of skill, know-how and technical experience, especially its experience in Tunisia if any.
(2)   the importance of the firm in financial terms: the value of its equipment, of its capital, and of past projects carried out by the firm.
(3)   the price proposed by the firm to carry out the project, and the eventual sources of credit that it could bring to the project.
(4)   the reputation of the firm in terms of its projects and its past relations with different clients, mainly in terms of conflict.

Qualified Tunisian firms are generally chosen in cases of international bidding if their prices are less than 20 per cent above their foreign competitors.

Examples of Projects Carried Out During the Sixth Plan Period (1982–6) with the Participation of Foreign Contractors
Since it is impossible to analyse all projects carried out in Tunisia during the sixth plan period (1982–6) in order to know exactly what was the degree of involvement of foreign firms, we shall limit ourselves to all projects of the Ministry of Public Works in three different areas: hydraulic engineering, road construction, and port construction. The total number of projects carried out between 1982 and 1986 was 28:

5 dams, 2 headwater channels, 2 commercial ports, 7 fishing ports, and 12 projects in different types of road construction. The approximate present value of these projects was about 850 million dinars. Two-thirds of this total investment was financed through public sources of finance. One third was financed through international sources of finance.

Thirty-four consulting firms were involved in the design stage of these projects (although this number involves double-counting for some firms with multiple projects). Seventeen consulting firms were Tunisian (mainly in road construction) and 17 were international, coming from France (10), mainly in port construction and hydraulic engineering; the USSR (3) mainly in dam-construction; West Germany (2) in hydraulic and road construction; China (1) in headwater channels; and Canada (1) in hydraulic engineering. The number of contractors who participated in carrying out all these projects is relatively high: 93, including all subcontractors. Since some participated in several projects, the actual number of contractors involved is lower. Fifty-eight contractors were Tunisian and 36 were foreign: France (12), Italy (6), Canada (4), Yugoslavia (3) West Germany (2), Greece (2), Bulgaria (2), China (1), Japan (1), The Netherlands (1), Spain (1), and Turkey (1).

We have noticed a high correlation between the source of finance and the nationalities of the consulting and building firms. This relationship is less evident when the project is financed through the World Bank or another multinational source.

Relations with the Tunisian Central Bank, Customs, and Fiscal Administration

An agreement between an international contractor and a Tunisian client is valid only after the approval of the Tunisian Central Bank. In general the global price of the project is given in both Tunisian dinars and in foreign convertible currency. This global price of the project includes study, construction, equipment, transportation, insurance, and training.

The international contractor is supposed to have two types of bank accounts: one in dinars and the second in a foreign currency. These two accounts can only be credited through transfers related to the project and no other source of transfer is allowed.

The first difficulty appears, before the project even starts, during the negotiations between the foreign contractor, the Tunisian client, and the Tunisian Central Bank in order to fix in advance which part of the budget should be in dinars and which part should be in foreign convertible currency. These parts are related to the sources of finance of the project and to the importance of subcontracting operations that could be done by Tunisian firms. The international contractor must spend in Tunisia and subcontract to other Tunisian firms an amount at least equal to the part of their project that is in dinars.

After Central Bank approval, the international contractor must get a permit to work in Tunisia valid for the period of the project from the Ministry of National Economy. Thanks to this permit, the international contractor is allowed to work as any Tunisian firm, to hire workers, to pay taxes, to have an official address in Tunisia, etc. Next, the international contractor has to get an identification number from the Customs Administration. Thanks to this number, the international contractor can import with temporary permits, without paying any duty, all the equipment he needs on the condition that he re-exports the equipment to his home country once the project is over. But, since customs duties decrease by an eighth every six months, if the project lasts 4 years, he may sell his equipment in Tunisia without paying duties. Often, when the project lasts more than two years, the international contractor prefers to sell his equipment in Tunisia, generally to the Tunisian client, instead of re-exporting it. This happens whenever the cost of transportation is equal or superior to the remaining customs duties. Let us note that the equipment imported to carry out a definite project cannot be used to carry out another project, unless the Customs Administration agrees.

As for taxes, the international contractor is a Tunisian firm as long as the project lasts: the international contractor is supposed to pay taxes on all activities in Tunisia: construction, transportation, consulting, training, etc. Only overhead costs are deducted from the estimate of the taxable profit. Long discussions are often held on these overhead costs. In general, the Fiscal Administration does not accept that overhead costs should go over 10 per cent of the value of the production in Tunisia.

Relations with Tunisian Subcontractors
Although the international contractor is generally the main contractor and is responsible for the entire project, he is supposed to subcontract to other Tunisian firms all parts of work or equipment supply that could be bought in Tunisia, at least for an amount equal to the part of the project that is financed in dinars. This obligation may be a source of problems to the contractors if some precautions are not taken in advance; especially when the contractor is supposed to help the Tunisian subcontractor technically and financially.

A list of all works that can be undertaken by Tunisian subcontractors is given in the appendices of contracts. This list must be followed except where:

(1) the construction of any other work could not be done by any Tunisian subcontractor (in this case a written justification is required);

(2) the local quality is not technically acceptable;

(3) the delivery timing of the subcontractor is not compatible with the whole project timing; or

(4) local prices are much higher than those of foreign competitors.

According to some Tunisian contractors, this obligation to subcontract is imposed by Tunisian clients less and less. The clients are mainly public agencies, under the pressure of the World Bank and other lenders who prefer to give more freedom to the main international contractors. Tunisian contractors, however, believe that they are technically as efficient as foreign contractors for most types of projects. If they do not win certain projects it is mainly because international contractors are supported by credit systems from their own home countries. They win contracts because of finance.

According to these Tunisian contractors, they are actually less expensive than foreign contractors in Tunisia: they pay the same wages to local labour, they get local materials at the same price, their Tunisian skilled labour is less expensive than imported foreign skilled labour, transportation costs for existing equipment are lower for Tunisian firms, amortization is equal for the same quality of equipment. However, international contractors appear to be less expensive in case of bidding because they have more credit facilities from their home country banks, don't import new equipment for their projects in Tunisia, and therefore they don't include amortization in their total bid. Since they dispose of convertible currency, they are more efficient at buying and importing spare parts, tools, and other materials or services necessary to their sites in Tunisia. Thanks to this last advantage, foreign contractors have the reputation of keeping to timing and the schedule, which is not the case with Tunisian firms. But this is not their fault: the Central Bank, the Customs Administration and the Fiscal Administration are responsible for the delays.

Some Tunisian contractors propose that foreign contractors be subject to a joint-venture requirement with Tunisian contractors which, they think, is much better than the present situation with most Tunisian firms being only subcontractors on big projects. These Tunisian contractors insist that if they are expected to export their services abroad they have to get more experience on large local projects; and this experience can only be obtained through joint-ventures with international contractors and not through subcontracting alone.

Labour Relations

Before the project starts, the international contractor draws up a list of all his needs in labour: unskilled and skilled workers, foremen, engineers, etc. This list is then discussed with the Tunisian client and the regional labour inspector. The maximum number of required workers has to be hired in the region where the project is going to be carried out. In case of shortages for certain types of skills, workers including foremen and engineers, could be hired in other regions of the country. However, the international contractor may have (and may insist upon) permission to bring with him a 'task force' composed of a certain number of key men, mainly foremen and engineers. This task force has to be discussed with the Tunisian Administration represented by the regional labour inspector and with the Tunisian Central Bank who prefer to have the minimum possible amount of wages and salaries transferred abroad.

Tunisian law (labour law and social law) has to be applied to all labour whatever their nationality. However, the workers belonging to the contractor's task force who were hired before the project in Tunisia started can benefit from the most favourable clause between Tunisian law and their home-country law. All labour must be insured with Tunisian insurance companies. Tunisian labour must be paid Tunisian wages (minimum wage regulations) while foreign labour is paid according to international levels. Wage-increases have to respect a specific formula that the international contractor and his Tunisian client agree on before the project starts. The contractor's task force and any other foreign worker or engineer must pay their income taxes in Tunisia and according to Tunisian law whenever they remain in Tunisia more than 183 days.

This fiscal regulation generally disturbs contractors who must either choose to change all their task force before they spend 183 days in Tunisia, a decision that could be harmful to the normal progress of work; or whenever the permanent presence of some skilled workers or foremen is required on the site, the contractor agrees to pay their income taxes in Tunisia. In this case, since income-tax rates are very high in Tunisia, they try, with the co-operation of Tunisian fiscal experts, to under-declare their incomes.

Very often, the international contractor must organize training of Tunisian engineers and foremen. The minimum number of training-months is generally fixed by the contract. This is the case of some projects where technological transfer is relatively high, such as power plant construction by a Japanese firm in the area of Tunis.

Other Requirements

A contract between an international firm and a Tunisian client generally contains a surprising number of clauses and articles referring to responsibility, control, conflict, transportation, patents, and insurance. According to some lawyers and other experts, these clauses seem to be too many when there is no conflict; but whenever a conflict appears between the contractor and the client, these clauses are not at all sufficient to solve the problem, despite their numbers.

(1) Clauses Concerning Transportation In most contracts we find that all transportation of equipment, materials, etc. from abroad or from other Tunisian cities to the site must be done under the responsibility of the contractor, and all costs are paid by him. Transportation must be done from abroad to Tunisia on Tunisian ships, unless it is impossible. Of course, as mentioned above, all trucks and cars necessary to the construction of the project are imported to Tunisia free from customs duties. But after the project is over, they have to be re-exported, unless they pay the remaining required duties. Anyway, these cars and trucks could not be used for any purpose other than the project itself.

(2) Clauses Concerning Insurance The contractor is supposed to pay all the required insurance. All these insurance contracts have to be made with Tunisian insurance companies. The amount of required insurance varies from one project to another, but generally we find the following sorts: on all types of transport on site and out of site; on civil and professional responsibility in order to cover all types of errors in the study phase or the construction phase; cross-liability; professional sickness and work accidents.

(3) Clauses Concerning Technology Transfer and Patents According to these clauses, the international contractor should not use a patented technology without a written permit of the owner of the patent. Anyway, in case of conflict, all payments have to be made by him. On the other hand, if the international contractor innovates during his work in Tunisia, the benefits of the new technology have to be shared with the Tunisian client. The relative shares of each have to be negotiated. Once the project is over, in order to get spare parts, priority is always given to the international contractor by the Tunisian client, for equivalent price and quality conditions.

(4) Clauses Concerning Liability, Control, and Conflict The international contractor is responsible for all parts of the work done directly by his own services or indirectly by different subcontractors. In case of a conflict between the international contractor and the Tunisian client, mediation and/or arbitration procedures exist. The typical procedure is as follows. First, a mediator is designated. The procedure for designation varies from one contract to another; in some he is designated by the president of the Tunisian Engineers Association. Second, if the mediator cannot solve the problem, the arbitration procedure is started. One arbitrator is designated by the contractor, another by the Tunisian client, and eventually a third is designated by the first two. If these procedures fail, the president of the Court of Justice of Tunis handles the conflict. Tunisian law is very often the only one that has to be taken into account. If, for specific points, Tunisian law is silent, international law (mainly French law) may be taken into account.

The causes of conflicts are numerous. Some start when the Tunisian client decides to fix penalties for delays by the contractor or for technical defects. Other conflicts start when the Tunisian Central Bank refuses to allow the Tunisian client to pay the international contractor for one reason or another. Others are generated by a misunderstanding of a fiscal clause or a customs rule.

Notice that a certain number of guarantee deposits are generally required of the international contractor by the Tunisian client: performance bond, deposits to cover penalties, guarantee deposit, etc. In a few cases work has stopped on the construction site and before the contractor had finished his project. In most other cases however, thanks to the intervention of specialists (engineers, fiscal advisors, lawyers, auditors, etc.) and political authorities of both countries, solutions have been found to conflicts.

3 The Recent Experience of Tunisian Contractors Abroad

In recent years some large Tunisian contractors have attempted to extend their activities abroad. The principal Tunisian contractors who have built abroad are Afrique Travaux, Bouchamaoui Frères, COMETRA, Entreprises Ali Mhenni, ETRAPH, INCO, Societé Tachid, STAM, and TED. Some attempts have been successful — mainly in Algeria and in Libya — but some others in the Middle East, mainly in oil exporting countries, were not successful because of the high level of competition there. Tunisian contractors believe it is impossible to underbid Koreans who treat their workers like low-paid soldiers.

The Tunisian experience of exporting construction services is relatively recent: since the mid-1970s in Libya and since the early 1980s in Algeria. The main characteristic of this experience is its dependence on political relations between Tunisia and these countries. When these relations are good, Tunisian contractors are welcome. On the contrary, when they are bad, Tunisian contractors are no longer accepted.

The Tunisian experience in exporting construction services was very limited before 1983, about 8 million dinars of income flows into the country. In 1983, the Algerian and the Tunisian governments signed a contract authorizing Tunisian contractors to build low-cost houses in the eastern regions of Algeria, not far from the Tunisian border. Between 1983 and 1986, about 10,000 houses were built by about 10 Tunisian contractors in 6 different cities. Thank to this contract, the income inflows to the country due to 'major works' increased from an average of 8 million dinars before 1983 to an average of 24 million dinars after 1983.

Although the main reason for the 1983 contract is political, it is important to note that it also has economic and social aspects. Compared with Algerian contractors, Tunisian contractors have long experience in social housing construction (20 to 30 years). Competition has always existed in Tunisia among local contractors; therefore Tunisian contractors have succeeded in controlling their costs and in improving the quality of their production. Large Tunisian contractors are competitive with international contractors either in Tunisia, Algeria, or Libya and can win contracts for important projects.

In addition, Tunisian contractors have adapted low-cost housing to the social, human and economic needs of Tunisian households in terms of architecture, urban planning, cost, etc. Since Algerian households are not very different from Tunisian ones in terms of income levels, family sizes and social behaviour, the Tunisian experience is relevant to Algerian needs, and Algerian contractors can learn from them.

Some contractors are very satisfied with their experience in Algeria because they could get large projects (600 dwellings per year) that they could never get in Tunisia (an average of 100 units per project). These contractors have more advantages from scale-economies in their projects in Algeria than in those which are located in Tunisia. One of these contractors even revealed that his average costs are lower in Algeria.

Of course, most of the contractors complain about the lack of mobility of capital, of equipment and of labour from one project to another in Algeria, and from the heavy bureaucracy that they have to deal with. They have noticed that since the fall of oil prices, the Algerian Central Bank authorities are less flexible on transfer operations. For example, three years ago a Tunisian skilled worker or foreman was allowed to remit 40 per cent of his income while he was working in Algeria, but since 1986, he could only remit 20 per cent of his income.

In general, Tunisian contractors are satisfied that they have a good reputation in Algeria and that they have advantages over other international contractors even if these advantages are not due to government support. The Tunisian government does help Tunisian contractors to extend their activities abroad, but these encouragements are still modest if they are compared with those that merchandise exporters get. Tunisian contractors exporting their services abroad could use more advantageous rates of interest, better tax advantages, and more flexibility in their dealings with the Central Bank. They need the same advantages as the merchandise exporters. After all, the export of construction services to Algeria has generated an increase in construction materials exports from Tunisia to Algeria.

11

Conclusion: Comparison and Analysis

In this last chapter we shall see what light the country chapters have shed on the issues raised in the introduction and will try to sort out a few contradictory implications. In order to present the great diversity of projects, firms, home and host country policies somewhat coherently, similar cases will be grouped in a way that shows the basic long-run development pattern of international construction. This pattern is one that arose in the nineteenth century and is likely to continue into the twenty-first. Dramatic events like wars and the OPEC oil crisis have intruded now and then, leaving permanent marks. The effectiveness of markets and government intervention has also been shaped and limited by the basic pattern, as will be reviewed later in the chapter. Special stress will go on finance and technology. Finally, we shall try to assess what the future holds for this industry.

In summarizing and analysing all this, we shall occasionally draw on data from countries other than those in our case studies. After all, the overseas construction of Britain, the Netherlands, India, Turkey, and Yugoslavia is in a class with several of the countries we have included. Canada, Australia, New Zealand, and Israel should also be included together with most other countries from both Western and Eastern Europe, plus a dozen or so developing countries, including Argentina, Mexico, Colombia, Kenya, Senegal, China, Taiwan, Pakistan, Singapore, Hong Kong, and the Philippines. We shall begin with the basic pattern.

1 Basic Pattern of Structures, Activities, and Countries

To keep ideas manageable, we may think in sets of three for the basic types of structures and the main activities that produce them. A preliminary summary is that all three types of structures are demanded by clients from all types of countries (in varying proportions), but that all three types of activities are not carried out by contractors from every type of country for every type of structure.

Structures
The three types of structures are (1) the small and simple, (2) the large but conventional, and (3) the technologically complex and novel. Retail stores, highways, and petrochemical plants illustrate the three types. As every author has confirmed, in Third-World countries the large or technologically complex structures are most likely to be built by foreign firms.

Activities
The three basic activities are (1) project initiation, (2) organization, and (3) site execution. The smaller and simpler the project is, the more importantly will site execution loom as the basic construction activity. Indeed when most of us first think of construction we picture men and machines digging foundations, assembling materials, and installing fixtures. But the cost of site execution depends greatly on organization: scheduling, subcontractor recruitment, mobilization of labour, materials procurement, quality control, and financial management. Once master builders or general contractors did the organizing as merely an aspect of site execution; but with more sophisticated record keeping and especially computer-based data systems, we have seen firms evolve into specialized 'construction managers' for large and complex projects.

Project initiation begins with feasibility studies and goes on to mobilizing finance, developing the design, and creating a consortium of firms, if necessary, for later organization and site execution. The case studies have shown that construction firms in every country believe that contracts are often obtained with this initiation work or not at all. Doing the feasibility study puts one a notch up for getting the design contract, and the integration of responsibility for design and site execution is increasingly commonplace. In the private sector some constructors operate in no other way. Moreover, in a buyers' market, we have seen that some large contractors now find work in capital-poor countries by approaching clients with both the idea for a feasible project and a scheme for financing it.

Countries
More elaborate schemes are possible, but we simply classify countries in two groups as (1) advanced or industrialized or (2) developing, 'newly-industrializing', or late-coming. From some advanced countries contractors ventured abroad as early as the nineteenth century and continued with such work afterwards. Thus the British and

French built railways, canals, harbours, and commercial facilities, not only in their empires, but throughout the world. The Germans built similar facilities in Russia, the Balkans, and the immense Turkish Empire. During the nineteenth century Americans were busy investing in their own continental expansion, and the Panama Canal may be seen as part of that. Early in the twentieth century, however, American constructors took the latest technology of the time abroad: hydroelectric power stations, petroleum refineries, and mass-production manufacturing plants.

Three of these countries are represented in our book. As we have seen, they kept winning contracts abroad between the World Wars despite the depression. After the Second World War the European firms quickly resumed their foreign (for a while, still mainly colonial) construction. American firms, with no diversion for postwar reconstruction at home, took the lead abroad during two prosperous decades, but as late as 1972 the largest 400 contractors still had less than 10 per cent of their work abroad.

Other advanced countries did not lag much behind in technological sophistication; but early in the century, they were either poorer or smaller and lacked construction firms that could venture abroad. Italy, Sweden, Denmark, and other European countries, as well as Japan and Canada belong in this category. In the contributions to this book we saw that Italian and Japanese overseas contractors were limited to their own empires before the Second World War. About a decade afterwards, the Italians began to exploit their Alpine hydroelectric expertise with projects in Africa, while the Japanese were still busy with reconstruction work in former occupied areas. Their push to go abroad competitively did not come until the late 1960s. Meanwhile, the Scandinavians built primarily for their neighbours, including one another, around the Baltic Sea and exported their prefabricated housing techniques further afield until the vogue for that was interrupted, perhaps temporarily.

The latecomers are countries substantially poorer than the advanced but are often classified as 'newly industrializing'. In this book we have chapters on three that illustrate the diversity of the category: Brazil, Korea, and Tunisia. A dozen other developing countries like India, Turkey, Kenya, Colombia, and the Philippines have also had contractors who have ventured abroad for work. Directly or indirectly the oil boom of the 1970s made their entry possible. As construction work around the Persian Gulf quintupled for foreigners from ten to fifty billion dollars annually and several hundred thousand construction workers migrated in, shortages and opportunities for builders appeared throughout the rest of the world. Now that the Middle Eastern volume has fallen below a third of its peak level, the latecomers throughout the world have been hurt much more badly in their international work than firms from advanced countries, as mentioned in Chapter 1. From 1981 to 1986, the value of contracts awarded to Koreans fell by 85 per cent, compared with 55 per cent for all, and those awarded to Brazilians fell by over 90 per cent to a mere $34 million.

Interaction and Growth

As in other industries, the basic long-run pattern of growth, importing, and exporting reflects changes in demand and comparative advantage or supply. As countries develop, they accumulate capital and acquire technological skills. To an initial set of simple structures a network of large-scale overhead facilities is added, followed by a set of technologically advanced installations. Domestic builders necessarily lack the experience for constructing anything so large-scale or advanced, hence foreigners are brought in to plan and organize the work. Since they must be paid in foreign exchange, the initial projects are likely to be related to export promotion. If there is an intensive export boom, a gold rush or an oil frenzy, the domestic building industry may not even keep up with the demand for simple structures, and foreigners can be hired to build them.

From the foreigners or from study abroad, domestic contractors, usually engineers, will first learn how to manage the sitework of large-scale conventional projects. They learn as employees, subcontractors, or pro-forma national affiliates. In the course of taking on some work on their own account, skills of organization − scheduling, procurement, and financial management − are acquired.

Ferchiou's account of Tunisian construction shows not only the sort of regulations that governments impose on international contractors but also the pressures from domestic contractors that are behind these regulations. To begin at all, foreigners need an impressive array of permits. Then they must be at least 20 per cent cheaper than their local competitors, even if there is open tendering. The extent to which they organize their activity in terms of foreign or domestic currency is strictly controlled. Finance that is provided by the host country must not be converted to foreign exchange for imports but must be used for local purchases or subcontracting. Equipment, materials, and personnel that are brought in are also subject to detailed controls. Insurance has to be national. In Tunisia the amount of overhead that can be subtracted from taxable profits is limited to 10 per cent of turnover. National contractors must be hired for anything they can build in time with acceptable quality and costs.

Whatever constraints are imposed on foreigners, national contractors will consider them insufficient. The foreigners' superior access to finance is considered an unfair advantage that should not be allowed to offset a variety of advantages of national contractors with respect to employment, labour costs and other factors. Especially resented is the reputation of foreigners for punctuality when this is due to their international procurement practices which avoid the bureaucratic delays that are inflicted on national contractors by the very government agencies that complain about delay. Above all, local firms wish to be joint venture partners, not just subcontractors, so that they can learn how decisions are made and replace the foreigners altogether except for the most complex technical work. According to our contributors, the Koreans passed this stage in the mid-1970s, the Brazilians somewhat later, and the Tunisians appear to be on the verge of so doing. Some of the world's poorest countries are far behind these three.

With the help of supportive legislation, growing firms will displace foreigners and on occasion will take on projects in neighbouring countries. Thus Brazilians built in Bolivia and Paraguay, Koreans in Thailand and Vietnam, Tunisians in Libya and Algeria, and Kenyans in Somalia and the Seychelles.

Capital and technology, however, remain relatively expensive and the latecomers will be weak at project initiation and the building of advanced installations. Firms in latecomer countries cannot mobilize local talent for producing advanced designs, or if they could, will not inspire clients with the confidence that the work will be as good as that of more experienced, highly reputed firms from vanguard countries. At the same time, lack of capital in national financial institutions deprives latecomer firms of an important means of bargaining for projects during adverse times. Only during boom times can firms from latecomer countries move into other countries, especially capital-rich LDCs, to produce their specialties, conventional infrastructure and buildings. They are rarely found building technological novelties in industrialized countries. Among exceptions were two Brazilian alcohol-from-sugar plants in the United States.

Countries in Transition
Latecomers turn into followers when they have acquired the technical and financial resources to compete with advanced country firms in developed countries (which made up about a third of the international construction market in the mid-1980s). When a country's firms can compete anywhere, they have become advanced.

In the mid-1980s, the Koreans appeared to be in such a transition. Still, finance was scarce, leaving the Middle East as the primary market. Even there, the Libyan Great Man-Made River was being dug by Dong Ah under American organization in the form of construction management by an affiliate of Brown & Root. Elsewhere the Koreans hired technologically advanced firms as subcontractors, and the Brazilians did the same. Once that relation would always have been the reverse. Korean overseas construction awards in 1985 were 5.8 per cent of the world's total, over four times as much as their 1.3 per cent share of design billings; but what is significant is that they had foreign design clients at all.

Meanwhile, the Japanese and Italians, as pointed out in Chapters 3 and 4, were still deficient in international design consulting and computerized management, compared with the Anglo-Saxons, but no doubt that gap would not last. In the 1980s the Japanese and Italians had few construction management contracts, while much of the foreign work of some major American firms (Bechtel, Parsons, etc.) was of that type.

In 1985 the Japanese share of international design billings, 6.2 per cent, was less than half their share of construction awards, 14.3 per cent. However, the Japanese were becoming the world's leading creditors; and using finance they combined with Italians to win the Bosporus Bridge contract in competition with a consortium of British-led firms. Lack of finance had kept the Italians back until the late 1970s and then required the elaborate 'triangulation' schemes described by Aldo Norsa. Among other countries, the Scandinavian countries were technologically as advanced as any, with a high reputation for consulting services. Their 1985 share of international design billings was 7.3 per cent, lagging only behind the 12.7 per cent share of the British and the 32.0 per cent share of the Americans. But the Scandinavian economies were too small to provide the finance for large-scale project initiation abroad.

Being in the lead makes for complacency and missed opportunities. In 1985 British, American, and Canadian design firms still had over half (52 per cent) of international billings, but they were less resolute than others in using design as part of a strategy for getting contracts. For example the British share in design billing was 12.7 per cent, compared with only 6.9 per cent in contract awards. The Dutch had a 6.0 per cent share in design and only a 1.6 per cent share in construction. The Americans still had a third of both design and construction markets, but much of the latter was simply construction management, using their organizational edge and leaving sitework details to others.

With respect to finance, governments of the most advanced countries just mentioned, together with the Germans, more or less left contractors to their own devices. Despite much official posturing, they provided only limited financial support to offset the more glaring subsidies given by competing countries. The French were the exception to such complacency in action, if not words, and parlayed a 6.6 per cent share in design billings into an 8.2 per cent share in contract awards.

Fate of the Latecomers

The domestic economies of latecomers or LDCs are insufficiently developed to provide a home market for construction that allows great technical, organizational, and financial capability to arise. Indeed, the lack of such capability makes it tempting for contractors from advanced countries to move in. Barriers to entry are therefore widespread in LDCs, from the Persian Gulf to Patagonia. Whenever possible, the World Bank 'slices and packages' large projects to make them accessible to LDC builders and if per capita product is below $400 annually, gives them a 7.5 per cent bid preference in addition. As development proceeds, the domestic market expands, and capital, technology, and organizational skills are acquired.

In the meantime, latecomers are the most vulnerable to international recessions. Let us assume that the category includes all non-European firms besides the Americans and Japanese. From 1982 to 1985 according to *ENR*, their international construction awards fell by 63.5 per cent; while that of advanced countries fell by only 35.6 per cent. Awards to the latecomers declined from $22.4 billion to $8.2 billion; while others fell from $100.7 billion to $64.8 billion. If the declines had been proportional, the latecomers would have had about $5 billion more in contract awards, and the others $5 billion less. What prevailed was the last-in/first-out principle.

Nevertheless, things did not simply return to what they would have been without the OPEC price increase and financial recycling to Latin America and a few other countries. The skills and capital that Koreans, Brazilians, Turks, and others gained from working in Saudi Arabia, Iraq, Libya, etc., were long-run additions to their economies, raising incomes, markets, and construction capabilities. It meant that they would in time move to advanced status earlier than otherwise. The next strong regional construction boom might do it. In fact, Korea might attain that category soon without another such stimulus.

It could well be that in the late 1980s Chinese overseas contracting for labour services and labour-intensive infrastructure could echo the Korean success story of the early 1980s. With its first appearance in 1984 among the *ENR* top 250 – $762 million (1982 dollars) in contract awards – the largest Chinese company, the China State Construction Engineering Company already ranked 21st in the world. It had only been organized in 1982, but its exports were already above those of the second-largest German firm Hochtief, and just below the largest Japanese firm Kumagai Gumi. For 1986 the company reported $303 million in foreign turnover and $250 million in new contract awards ($237 million in 1982 dollars according to *ENR*, 16 July 1987;

domestic work was $377 million). It was the largest of 67 Chinese overseas construction and labour-service companies and did about one-third of their total overseas construction. At home its labour force was 1.48 million, so only a minute fraction, 13,000 workers, were building hotels, housing units, office buildings, and water-control projects, mainly in the Middle East, North Africa, and Asia. Altogether 170,000 Chinese were working abroad on industrial and construction projects (*China Daily*, 9 June 1986 and 27 February 1987).

The Chinese built not only in the relatively prosperous markets just mentioned, but also in Sub-Sahara Africa and Latin America, 50 countries in all. Loans to these countries were called 'interest-free', but to ascertain if that was truly the case, one would have to know Chinese costs. If extremely low costs are marked up enough to begin with, one can call the repayment stream 'interest-free' and still earn a good return on capital.

In developing countries the Chinese contracts were small but showed that the market was not closed to outsiders, as some have concluded on the basis of the Korean absence from many markets. For example, the China Harbours Engineering Company outbid Italians, Swedes, and Spaniards in winning an $8 million contract for rebuilding the harbour of Santa Marta, Colombia. The average size of its 13 Colombian contracts was less than $2 million. Most of its work, however, remained in the Middle East. The company had 4,000 workers abroad. In Africa the Chinese built in Mali, Burundi, Gabon, Nigeria, Tanzania, Zaire, and Zambia (*China Daily Business Weekly*, 3 September 1986; 25 February 1987).

2   The Extent of Market Competition

The basic long-run pattern of change in the international construction market, as just presented, fits a competitive model. But how competitive is this market really?

One possible view is that a single international construction market does not exist at all but only highly stratified and segmented sub-markets. Not only are there exclusive deals and set-asides, but the source of finance can be more crucial than the sheer cost of earthmoving and erection. Better financed contractors from rich countries can win out against lower-cost builders from developing countries because finance and construction capability are not kept apart. International open competitive contracting for multilaterally financed projects is a small percentage of work. More important are projects financed with tied export credits, tied foreign aid, or privately financed direct foreign investment. Price competition is limited in any case because of the difficulty of knowing costs in advance, especially on all those projects without complete preliminary (hence rigid) plans (that is, on all design-build and some negotiated contracts). The rationality of any markup for assured quality because of trust or the reputation of a contractor (intangible capital) is particularly hard to assess and to distinguish from mere prejudice and personal contacts.

Do all these practices not provide a buffer for distortions, inefficiency, and monopoly profits ('rents')? Do the results change who builds where and for what price and give a pattern different from that implied by perfect competition? In this section the evidence and logic for and against the prevalence of competition, as presented in the country chapters, is reviewed. The importance of coping with the great risks of construction is stressed. Normal profits can be defined only in this context. Risks are usually highest for new entrants, and we shall see that they foster specialization, integration, or diversification. Price determination is a key element of competition, and our evidence about that process concludes the section.

The Evidence for Competition
Construction is an industry that is easily entered. It is also characterized by the co-existence of a large number of firms. Those factors suggest a competitive industry. The great importance of mobile human capital makes it easily entered with thousands of firms ready to move from one place to another, expanding, shrinking, or learning new things simply by hiring and firing. Product characteristics change with each site so that economies of scale are limited, and all one can advertise is past competence – not an insuperable entry barrier.

For a large country like the United States, the concentration ratio as an index of competition abroad was found to be about the same as the ratio at home, not high. The largest four firms had 21 per cent of the work. For every additional foreign competitor abroad, an equivalent domestic rival seemed to stick only to the American market. For small and medium-sized countries, however, the international market was more competitive than the one at home (Seymour, 1986, p. 97).

The country chapters have indeed reported much evidence of competition. As the Middle Eastern demand quintupled during the 1970s, German, Japanese, and Korean firms came in and easily expanded their exports of construction services by a factor of ten, while that of the Americans just about quadrupled. In absolute terms their $35 billion increase was still more than the combined $30 billion rise of the other three countries. Despite all this expansion, contractors could not write their own ticket. In a survey of twenty overseas construction firms, Seymour found that all said their Middle Eastern clients bargained hard to get prices down (Seymour, 1986, p. 146). Once competitive bids were in, governments of countries like Iraq or the United Arab Emirates would select a few of the lowest for rebidding or for negotiating the price downward. Competitive bidding remained the most common way of awarding contracts (Demacopoulos and Moavenzadeh, 1985, pp. 73–4).

The many losses and failures of the 1980s are also evidence of intense competition to enter and to remain in the market. Perhaps the most spectacular collapse, among many, was that of the C. T. Main Corporation of Boston, which rose to first in the world in 1982, then met disaster and wound up as a subsidiary of Parsons a mere three years later. Germans, Japanese, and Koreans all made losses on their initial ventures into new markets; and sometimes bid on projects with over 20 competitors. Forty-five contractors bid on one World Bank project in Pakistan.

Koreans were accused of 'cut-throat' bidding as they moved into the Middle East in the 1970s, and they themselves accused the Japanese of such practices as that country shifted its stress to South East Asia in the early 1980s. The Koreans compared the winning Japanese bids with their own for 17 cases during 1981 and 1982. The Japanese were not just the normal 5–10 ten per cent low as winners, but anywhere from 20 to 75 per cent below the Koreans. Several Korean firms went bankrupt. During 1982–6 the Japanese share in the South East Asian market continued to rise from 23.8 to 31.2 per cent, while the Korean portion fell from 10.2 to 2.1 per cent in 1985. But from 1985 to 1986 the volume of Korean work in South and East Asia doubled to 4.9 per cent. The President of Takenaka Komuten, one of Japan's six construction giants said, 'To tell the truth, we've given up expecting to win orders for construction works, say, in Singapore or other South Asian markets in international open bidding . . . We now know how strong Korea, Inc. is.' After constructing the first phase of Singapore's Changi International Airport, Takenaka Komuten was replaced by Hyundai for phase two (Toichi Takenaka quoted in *The Japan Times*, 25 August 1986). In correspondence with the editors of this book, Mr Takenaka clarified that while he no longer expected to win South Asian contracts based on price competition alone, he believed that Takenaka Komuten was still competitive when technology, quality, punctuality, and financing were important factors.

Not only have contractors from every country demanded support from their governments against 'unfair' foreign competition, but executives (and observers) of American, British, Brazilian, French, Japanese, and Korean firms all said their compatriots were 'fiercely' competitive among one another. Such statements are neither dishonest nor compelling. Executives tend to consider competition too harsh any time they lose a contract. Some, however, claimed that contracts had been lost only because the winner had allowed nothing for profits and overheads and hoped to make it up on high charges for the inevitable changes in the design.

Competitive advantages do not determine who enters a wildly expanding market such as the Middle East during the 1970s. Competitive advantages determine who survives when that market contracts. In this diversified industry, the survivors will be those most efficient in particular types of project or construction activities. One would expect that efficiency to be primarily due to demand and supply factors in a firm's home construction market. In the mid-1980s, that sort of shake-out process has in fact taken place. Americans now typically win contracts for management and the installation of advanced process plants; the Japanese develop capital-intensive projects and dig tunnels; Italians design and build modern factories; Scandinavians promote innovative housing systems; and Koreans and Chinese do labour-intensive buildings and infrastructure.

For the client, either a low price or cheap finance mean lower debt service charges. They are interchangeable forms of monetary competition. One can even say that lower interest is a disguised form of price competition, nowadays perhaps its most prominent type, but we leave that to the later section on finance.

Evidence against Competition

Some contracts are awarded without any competition at all, but *whenever* one firm is kept from offering a lower-priced deal or another is given a special preference, competition is impaired. Evidence of such exclusions or preferences can be found in each of our country chapters and in other sources. Exclusions can be open or tacit, public or private, and be applicable to one's own or other nationalities. In Korea, for example, the Ministry of Construction decides which single firm can compete for any overseas project. Contractors, bankers, and government officials meet monthly, not only in Seoul, but also at the Korean Embassy in Saudi Arabia to co-ordinate bidding strategy, building plans, and exchanges of equipment and personnel.

More common official distortions of competition are those which limit the home market to domestic firms or joint ventures and those which help national construction firms abroad with special insurance, cheap credit, subsidies on building materials, and outright grants, often by way of the client. As Moslems, the Turks have the only major foreign firms allowed to build in Mekka. Countries that build bases and embassies abroad or grant foreign aid, like the United States, typically reserve the work to their own contractors or give them a hefty bid preference (20 per cent for the US in the Pacific).

In one country, identical power stations in close proximity were financed by the World Bank and by the United States. The US-financed station had to be built by US contractors, and (according to an official who was involved) the lowest bid was 15 per cent higher than the lowest bid for the Bank project, open to all.

The Middle East Division of the US Army Corps of Engineers gives an example of favouring one's own contractors rather modestly. The Corps has managed over $20 billion of construction programmes for the Saudi Arabian and other governments. Once designs were complete, the Corps usually furnished a list of about two dozen qualified contractors. US contractors complained that the list was not exclusively American. According to Congressional testimony, however, for designs the Corps used 'entirely US architect-engineer firms due to our insistence on US standards and specifications [which] consequently entails use of American-made equipment and materials' and gave American contractors an advantage (Testimony of 25 June 1979, House of Representatives, Committee on Foreign Affairs, p. 4, also quoted in Zahlan, 1983, pp. 110, 115).

In a worse case, the US was alleged to have pressured the government of Taiwan to shift contract awards for a mass transit system from a British to an American consortium after the British had already carried out preliminary work on the project for 20 months during 1985–6. The Americans called attention to the great trade surplus that Taiwan had with the United States and to the fact that the British (like the Americans) had withdrawn recognition from the government of Taiwan (*Construction News*, 12 March 1987).

Not even the regulations of the European Common Market have done much to overcome the 'procedural barriers and logistical obstacles' to internal competition from one member country to another, according to our Italian contributor, Norsa. Riedel's paper confirms that except for some projects near borders, little intra-EEC work goes on. The new branch of Philipp Holzmann in Italy is considered an anomaly. That more is possible, however, is shown by the Scandinavians. The best way to move in elsewhere is through a consortium that divides up the work. Thus Finnish contractors deal with the Soviet Union mainly through Finn Stroi in accordance with five-year trade agreements. There may be rivalry but no competition in the true sense.

In the private sector, negotiated contracts with favoured building firms are common. Multinational manufacturing enterprises establishing a new plant abroad commonly reserve the work to a compatriot contractor with whom they have had past experience. Some negotiated fiercely, other less so, but no information is available on the extent of either case. In any event entry is not open to all. Even more restrictive are contractors with a patented or 'proprietary' technology, like the M. W. Kellogg Company of Houston in petrochemicals or the Italian Impianti for seamless steel pipes. They seem to get all the contracts for certain types of installations, suggesting virtual monopolies.

Among the chieftains of the Arab World and some dictators elsewhere, the difference between the public and private sectors is often blurred. Negotiated contracts for public works via joint ventures with either a state-owned contractor or a building firm owned by a public official are said to earn exceptional profits. Firms willing to collect their share are praised for their rapport with the local culture. American and Brazilian firms were not the only ones surveyed here who have reported a number of cases. Obtaining such work and other favours depends on preferences through personal contacts. Every contributor has mentioned the importance of those. In their absence, one may have to pay 'fixers' who can provide them and 'skimmers' who could otherwise spoil them, as Said Aburish has described in vivid detail (Aburish, 1985).

Reconciling the Evidence

One cannot label competition either a fact or a myth by simply finding one of the two sets of evidence false. There are indeed many firms throughout the world who are eager to expand and who will attempt to enter new markets with offers that other firms see as disturbingly low. Yet the widespread existence of barriers to entry and inside deals is also obvious. Since we are not dealing with one market, but a great diversity of stratified submarkets simple generalizations are impossible. Wherever we can, we must analyse the effects of barriers to entry on the quantity, quality, and price of output.

To begin, one must distinguish preferences for a nationality from those for one or a few specific firms. In one type of case, a government action in favour of its nation's construction firms abroad may involve subsidized loans and tied aid costly to its citizens and prejudicial to other firms but actually lowering the price (repayment cost) to foreign clients.

If the client's country can be led to give the preference, then firms of the chosen nationality can co-operate to keep prices high, as the Finns do in the Soviet Union. Obviously protective measures that exclude all foreigners from a market will raise costs and prices until the local construction industry has developed enough in quantity and efficiency. But that is not the issue here.

Perhaps the most objectionable interferences with the market are those raising profits of a few selected firms. The assumption is that these profits are arbitrary and serve no social function. Prices charged by construction firms should be no more than the minimum needed to attract all the factors of production – labour, materials, subcontractors, capital, and management – to the work. Any higher price yields excess profits or 'rents'. Ideally, to measure the extent of monopoly and its rent, one would have to know the supply and demand conditions relevant to any construction project and then estimate the marginal cost and its difference from price. Since multinational construction firms, like other multinationals, engage in transfer pricing to minimize their global tax liability and to hide profits from competitors, accurate information is not likely to be available.

Also in a risky business, the returns on one project must compensate for the losses in others; or in anticipation, for the expected chance of loss. Since few industries are as risky as construction, prices must be raised to a level that allows for the chance of loss, especially when something novel is attempted, at times with special guarantees or exclusive arrangements. Novelty implies the unknown, hence risk.

Risk

The special risks of construction include those of the client, who may fear that costs will escalate unpredictably, that structures will be faulty and need frequent repairs, or that they will simply be abandoned, partly paid for but incomplete and useless. These fears have led to the tradition of inviolable prior designs, competitive tendering, bid bonds, performance bonds, and other measures that shift risks to the contractor.

Among the contractor's risks are inclement weather, delays in site availability, unforeseen subsoil conditions, inadequate detail drawings, late material deliveries, unanticipated price changes, faulty subcontracting, unproductive labour, strikes, political disturbances, getting paid on time, foreign-exchange rate fluctuations, transfer restrictions, and expropriation. The last three risks, especially, make foreign contracting more hazardous than working at home.

Higher risks lead to more careful and costly planning and bidding, so that small projects cease to be worthwhile. At home or abroad, each project is likely to be a large part of annual turnover, so risks cannot easily be spread.

Managing these problems involves skills at organization, rather than site execution, as discussed at the beginning of this chapter. Edmonds and Miles (1984, pp. 43, 48) have pointed out that uncertainty forces the simple builder to become a risk manager. Or in the words of a World Bank report (1984, p. 5), 'the central issue in the development of a country's construction industry is the growth of the human capacity to manage risks.' Inability to manage risks and a related reputation for untrustworthiness among local builders has provided much of the scope for importing construction services in developing countries. Foreign contractors may be interested in coming because of a current or expected slump in their home market, but they will necessarily encounter new risks when first entering a strange country where language, culture, and legal systems are unfamiliar. For a time, feared losses must be offset by special preferences or unusual gains.

When firms enter a new geographical area or start providing novel structures, the inducement that they sometimes have to obtain beyond those common to their nationality or just plain corruption can be seen as compensation for bearing the special risks of entry. Where technological innovation is entailed, the risks of research are also covered by the hope of monopoly profits through the patent system or from keeping proprietary secrets. As the construction industry in an area matures and clients become more sophisticated, one would expect these preferences and related profits to fall. The initial firms may continue to dominate, of course, especially if they encourage local participation. The reputation and the trust of local clients that they have obtained in a series of projects may be called an intangible asset, and they can earn an income from that. Shifting to other firms would mean transaction costs for the client. Obviously, this discussion applies to those widespread cases where open competitive tendering has not been mandated by governments or international financial institutions.

Can it be proven that inside deals (not counting those in blatantly corrupt societies) usually tie in with risks or have negligible importance? Incidentally, utopian levels of virtue were not reported anywhere by our case studies, yet outright corruption flourishes best where cheaper competitors have no chance of entry. Risks keep the cheapest price from being the best choice, but payments for reliability, quality, confidence, and reputation are not easily distinguished from unfair inside deals.

Favourite Suppliers and Prices
For competition to be effective, it is not necessary that every firm be willing and able to bid on every project, giving unique solutions for costs and prices. Brazilians need not compete with Americans for petrochemical plants, and Indians need not compete with the French for nuclear power stations. If the Americans have French competition for high technology plants, and Brazilians compete with Indians for railways and water works, the market will function. Effective competition means that enough viable and serious competitors have access to the client so that price offers fall within a rather narrow 'relevant

range.' Within that range, clients may persistently choose a favourite contractor from a preferred nationality, seeming thereby to confirm the notion that competition is a myth. But competition need not be defined so narrowly. It is not the force that determines who gets the contract (competitive tendering aside), but the force that sets a constraint to the price that may be charged.

The point was clearly brought out in Drewer's chapter. Scandinavian firms have a variety of 'routes of entry,' a unique image, socially oriented innovativeness, and association with a foreign policy that strikes a responsive chord among clients in developing countries. Nevertheless, 'for any given international project they still have to "pare their margins" and take an increased risk to win the contract, the same as contractors from any other country'. The Tunisian chapter by Ferchiou similarly reports that the welcome to Tunisian contractors in the Arab world had 'the main characteristic of . . . dependence on political relations'. Nevertheless, entry to the Middle East failed 'because of the high level of competition there'. As another example, according to Aburish, a major Swedish firm paid an $18 million 'commission' to the Governor of the City of Jeddah to win a harbour-building contract against a lower bidding Korean competitor. But to be in that position at all, the Swedish tender had to be lower than that of 13 other competitors (Aburish, 1985, pp. 48–9). In this divided market there are thus sections where the threat of competition had an effect on price levels. In other sections of the market tied arrangements of the various sorts described above prevailed.

As an afterthought, could it be that paying bribes is a type of price competition? Usually bribes are thought of as a cost of doing business, paying a commission for the service of arranging vital contacts or for protection. A charge for these services by the government (the usual client for construction) is a charge for something that it normally provides free to any obedient taxpayer. Hence some bribes may simply be secret ways of getting the contractors' prices and profits down. For the contractors such secret price competition avoids both antagonizing and informing competitors. Hence, the secret payment is in the interest of both parties. Much depends on identifying who ultimately bears the cost of the bribe and who reaps the benefits, not the easiest research task. (We are indebted to Elias Dinopoulos for this challenging idea.)

3 Business Strategy

The country chapters of this book highlight the strategies used by hundreds of international construction enterprises. The extent of competition and the characteristics of the basic development pattern reflect these strategies. They involve decisions about when to enter which country, what to export there, in association with whom, under what contract form, at what price, and with whose finance. Later come decisions about the inevitable declines in markets: specializing, diversifying, down-sizing or withdrawing altogether. The decisions are based on a firm's resources of staff and capital: labour, skills, equipment, finance, and intangible capital, above all, its reputation and capacity for risk management. The client, of course, thinks he is buying a structure, not importing construction services although that, in fact, is what goes on.

Entry

Entry begins with types of activities for which the firm has the most complete set of resources and with familiar types of project. Risks are likely to be minimized, as described in the previous section, for example, by accepting initial work for one's own government abroad or for a compatriot multinational enterprise. Firms from the advanced countries generally began overseas work like that. Wholly owned affiliates were set up abroad to manage the work, and these often went on to win new contracts from local clients. Among American firms, the Austin Company, Morrison Knudsen, Ebasco, and Guy F. Atkinson illustrate this approach. Forty per cent of the firms in Seymour's international sample entered foreign markets to undertake work for home country clients (Seymour, 1986, pp. 163–4).

Firms from follower or latecomer countries adapted their strategies to the conditions following the Second World War. The Japanese and Koreans learned modern construction methods from work for the American military at home and went on to other markets with co-ordinated support from their own governments or large trading companies. Scandinavians and Italians tied many of their initial ventures to post-colonial programmes for multinational lending, foreign aid, or their favoured standing with East European countries. Without such assured customers, firms would wait for three to five years before carrying out decisions to go abroad, as the Brazilians did, and then begin on a small scale, expecting and realizing a few losses. Middle Eastern clients willing to pay 10 to 20 per cent cash advances brought several firms over the entry hump.

Specialization and Finding Associates

To initiate, organize, and execute work abroad, a construction firm will often associate itself with others in a joint venture or consortium. Such association, as reported in our case studies face a dilemma: on the one hand, profits are charged as a markup on inputs and fall when associates provide them; yet on the other hand, some contracts cannot be won or executed without associates. The decision becomes complex when it risks loss of the very internalization economies that may have justified becoming multinational in the first place (Buckley, 1983; Casson, 1979; Caves, 1982; Dunning, 1981; Hymer, 1976; Rugman, 1981; Williamson, 1985). In many developing countries local associates are legally necessary though exceptions can be made for work financed by the World Bank and similar institutions. At times associates are also needed because of the scale of the project or because a firm cannot provide some components at all and others not efficiently.

Having a Japanese or French associate, for example, gives access to cheap finance and credit for materials procurement. This motive for joint-venturing is new. Neo did not find it in the mid-1970s though even then clients frequently asked potential contractors to find export credits and loans for downpayments (Neo, 1975, pp. 112, 237, 253). Joint ventures with Korean, Chinese, Thai, and Philippine contractors have provided cheap and disciplined site labour for distant areas. As latecomer countries moved from sub- to prime contracting, many firms from advanced countries found it profitable to provide technologically complex components as subcontractors. For example, American firms have been subcontractors to Koreans in Saudi Arabia helping with components for cement plants, solar energy installations, and with an unsaturated polyester resin plant (Westphal et al., 1984, pp. 525, 529).

When it is necessary to participate in a consortium, firms prefer to be the one with a leading role. The standard way of having associates in a subordinate role is to be the general contractor giving out subcontracts. One keeps the main responsibility but loses some control. An American construction manager told one of the editors that he selected general contractors by asking first, 'show me your subcontractors'. Italian firms, by contrast with the Americans, insisted on direct responsibility for sitework in order to retain a better grip on quality and to avoid loss of know-how. Of course, Italian intermediate managers could be hired and transported at substantially lower salaries than Americans.

 The strategy of some American firms was to sell their general management specialty with less risk through the device of construction management contracts. But less than 5 per cent of the overseas American CM work was on a fee basis with little risk or liability on the project. On the remainder, risk exposure was over $500,000 or similar to that of a general contractor (ENR, 17 April 1986, pp. 92, 98–9). The principal Italian and Japanese construction firms rarely or never made such contracts.

Labour and Markups

Abdication from direct sitework was not the preferred strategy for many firms because it limited opportunities for generating profits. Where possible, companies provided all the labour force and materials, as the Scandinavian conglomerates Finn Stroi and Skanska did in Eastern Europe but not in the competitive arena elsewhere. During the heyday of the Middle Eastern construction boom, American firms often hired European and Far Eastern technicians and billed for them at much higher American salary levels, putting markups on top of that.

With stringent competition, however, all contractors have to turn toward the cheapest effective labour supply and try to keep markups a shade below that expected of others. This need favoured Korean firms because they could employ labour of their own nationality more productively than non-Koreans could. They needed less translating and less intermediate supervision, but had more discipline, and wages as much as 40 per cent below what foreigners would pay these same workers. This disparity, however, led to a violent strike, murder of a Korean supervisor, and summary execution of three workers by the Saudi police as a warning to troublemakers (Aburish, 1985, pp. 78–9). Koreans found that when they employed workers of other nationalities, efficiency was only 80 per cent as high as that of their own workers. According to our Korean contributor, Dae Chang, around 1980 a fifth of Korean workers abroad were employed by foreigners, while foreign labour made up over a quarter of that employed by Korean overseas firms. In the Middle East the number of Koreans rose from 6,600 in 1975 to 150,000 in 1983.

The Brazilian contractor, Mendes Junior, expected to have the same advantage of an integrated national operation as the Koreans did by transporting 12,000 Brazilians to the railway project in Iraq. As it turned out however, Brazilian workers abroad were found to be neither that cheap nor that productive; and after bringing in 6,000, Mendes Junior turned to Chinese labour contractors for the rest.

Competition, Conglomerates, and Finance

In general, association with other firms does go with specialization, a condition brought about through the pressures of competition. Willingness to specialize often helped to preserve one's market share in the early 1980s. The Italians, for example, made factory building their dominant activity, using knowledge so specialized that it could not easily be transferred, according to Norsa. As mentioned before, they tended to avoid subcontracting. At home Italians not only had a high degree of vertical integration in conglomerates with suppliers and clients, but Italian firms also set up consortia with one another, so permanent that some, like Impregilo (Impresit, Girola, and Lodigiani) became normal firms. In all these respects, the German experience was quite different: little specialization in factory building or anything else, little vertical integration at home, and a falling international market share. The Italians' functional specialization meant that they had to be geographically much more diversified, doing as much building in Latin America as in the Middle East during the peak of the boom while the Germans confined themselves to a few oil exporters.

Becoming a conglomerate in one's home country sometimes means that construction companies are bought by others as financial moves with little operational significance. In the United States we thus find Badger acquired by Raytheon, Brown & Root by Halliburton, M. W. Kellogg first by Wheelabrator-Frye and then by Allied Signal, and so forth. None of these mergers brought contractors together with typical clients, equipment makers, or material suppliers.

Firms in developing countries also formed conglomerates. The three main Brazilian international contractors all used their success to organize ventures in related manufacturing, mining, transport, and trading enterprises. The Korean firms who best survived the dramatic fall in demand, Hyundai, Dae-Woo, and Samsung, were the ones who had diversified away from sitework construction.

Both Korean and Brazilian firms had to take their latecomer deficiency in finance and designing skills, and their reputation into account in choosing associates. Korean firms began as subcontractors who supplied labour or a standard civil engineering capability to vanguard-country firms. Later, as already mentioned, they became the prime contractors who subcontracted technological specialties to advanced-country firms, or they participated in consortia because of lack of finance. Korean firms were typically under-capitalized with a high debt-equity ratio, mediocre creditworthiness, and with serious cash-flow problems. Much of their revenues had to be paid to service debts with high interest rates. As their Arab clients shifted from advance deposits to payment lags worth billions, those Koreans who did not go bankrupt, naturally had to raise their bids to cover higher risks and financing costs. The number of joint ventures rose from three in 1981 to eight in 1984.

German and American firms tried to slow declines, not by expanding contracting into risky ventures, but by undertaking ever smaller projects at the expense of minor competitors, usually compatriots. ('The Big Builders Learn to Think Small: Mega-projects are Scarce, so Bechtel, Fluor and Other Giants are Chasing Jobs They Would Have Spurned Before,' wrote Thomas Hayes in the *New York Times*, 28 July 1985.)

Japanese construction firms undertook real estate development with their extra capital in Australia and the United States and preferred to work with, rather than threaten, local building firms. 70 per cent of $1.8 billion in Japanese construction in the United States in 1985 was done for Japanese clients, using American subcontractors, and most of the rest went into residential and commercial property development (Pinyan, 1986, p. 13).

Preferred Contract Types
Construction firms seek the maximum amount of work and income together with minimal uncertainty. They see large but safe turnover as the key to profits. They want to initiate, organize, design, and build one project after another with no rivals muscling in. Most continuity, however, comes with being the design-build division of a corporation or government agency, a conceivably unattractive subordinate position. Such divisions like to be spun off so that they can independently pursue additional work on their own. After that has been obtained, however, with an appropriate expansion of the staff, firms seek to have as much of a commitment as possible for future work, if possible for decades ahead, in order to avoid the hazards of discontinuity. They dream about all the innovations that they might introduce if only continuity of work were assured.

Short of such a utopian development, the rank order of preference of firms toward types of contracts is very clear. Design-build package deals are usually most preferred by established firms, and open competitive bidding least. Design-build allows the most work for a percentage markup, the internalization of transactions, and the least worry about unforeseen costs, especially with cost-plus arrangements. These costs merely need to be reasonable so that one contract leads to another and so that one's reputation (or intangible capital) will be enhanced. The Italian research centre, Nomisma, estimated the amount of 'repatriable income' to be 12–15 per cent of contract value for design-build turnkey projects, but only 2–3 per cent for other projects. These levels match returns reported in *ENR* for the upper and lower quartiles of Italian firms in 1980 and 1981 (*ENR*, 16 July 1981 and 15 July 1980).

Design-build with a maximum price for the client introduces an element of uncertainty about profits that must be covered by higher contingency charges. Negotiated contracts raise the chance of some other shortlisted firm getting the work, but may allow some favourable modification of the design. To be shortlisted and chosen for any of these contract types requires the presentation of rather elaborate proposals. They do not always require the precision and care of competitive bid offers but can still cost as much as a 0.5–1 per cent of final project costs.

Competitive bidding, whether on open or selected tender, is least preferred by established firms because of the relative rigidity of both design and price and because it gives latecomers like Koreans, Brazilians, and Indians a chance to enter. To obtain rigidity, lowering uncertainty for the client and his creditor, is of course why competitive bidding was developed in the first place. It puts the burden of risk on the construction firm, a condition that the strongest firms from advanced countries try to avoid. Latecomers like the Koreans tend to lack a reputation for design skills that allow them to undertake turnkey projects, and so most of their work is likely to be on competitively bid standard projects with government clients. Like the Brazilians, however, they realize that lack of a design capability reduces knowledge of work being initiated and raises the chance of interference and dependency on foreigners. Japanese firms will also bid competitively when moving into a new area or when something grand like the Bosporus Bridge is at stake; but according to our chapter by Hippo and Tamura, they consider their high share of design-built work a testimonial to their capability and prefer that. Most contractors of every nationality agreed, but a few firms like the American Guy F. Atkinson and Perini prided themselves on being strictly hands-on contractors.

Bidding Strategy

If competitive bidding cannot be avoided, the essence of strategy for that is the same as for any other part of the business: spend whatever brings net gains but no more than that.

An adequately studied, estimated, and presented bid will cost between a 0.5 and 1 per cent of the final price of a wide variety of large complex projects common in international construction. This bidding expenditure must be related to the profits that will result from various percentage markups and the odds of winning the contract. These odds depend on the expected behaviour of competitors. Experienced contractors know who their competitors usually are and develop a data base of past bidding patterns. They also try to keep track of the state of the order books of competitors and whether or not they are likely to benefit from special preferences from particular clients.

During the 1980s *ENR* reported the median level of international profits on turnover as being 4–5 per cent, with the average level about 1 per cent above the median. Firms in the upper quartile reported profits of 7 to 10 per cent of turnover (see Table 11.1). The level of profits on projects worth more than $50 million was half that of smaller projects, but the rate of return on capital may have been the same due to economies of scale. Turnover rises compared with capital (Hillebrandt, 1985, p. 148). For the early 1980s, Zahlan (1983, p. 118) has also reported a 5 per cent level of contract earnings in the Arab world.

If nothing is known about one's competitors, one may assume that they have an even chance of winning any given contract. Consequently, if bidding costs 1 per cent of the final price and expected profits are 4 per cent, then one is likely to break even or better only if one limits bidding to projects with four or fewer serious competitors. The expectation is that one out of four will be won. The calculation is somewhat more complex than dividing profits by bidding costs since profits are in the future and must be discounted.

Table 11.1   *Profits as a Percentage of Turnover, 1980–86*

Year	Number of firms reporting	Median (%)	Upper quartile (%)	Lower quartile (%)
1980	268	5.0	9.9	2.9
1981	324	3.7	7.4	0.0
1982	80	5.0	7.0	3.0
1983	67	4.0	7.0	2.0
1985	50	4.0	10.7	1.3
1986	48	4.4	7.0	3.0

Source: ENR, 16 July 1981; 15 July 1982; 21 July 1983; 19 July 1984; 17 July 1986, and 16 July 1987. The figures for 1984 were not reported.

Note: The lower rate of response for 1982–6 could mean that median and average profits were actually lower because non-responding firms did less well.

Bidding strategy among the firms studied for this book, when reported, appear to fit that pattern. Korean firms were new to the market as prime contractors in the late 1970s and had no data about the past bidding patterns and strategies of their competitors. Hence they based their bids on estimated costs alone, according to Chang's contribution. As it was, their strategy resembled that of firms with decades of overseas experience. During the good times of 1980, they won 36 per cent of competitions and reported a median profit level of 8 per cent, almost as much as the American 10 per cent. As competition stiffened with the market decline, the Koreans, like the Americans, faced more competitors for each job, and the proportion of contracts won accordingly fell to 23 per cent in 1984. The winner was usually less than ten per cent below the second lowest bid. Note that for large contracts in the mid-1970s, according to Neo, competition was not likely to exceed two bidding consortia, raising the expectation of profits (Neo, 1975, p. 117).

As reported in Chapter 2, the largest firms do approach bidding strategy very systematically. Top management authorizes an annual bidding plan with a budget for as many as 50 employees in the estimating department. Companies do not bid in response to every solicitation; but when they do, resources are allocated in proportion to the expected present value of additional profits and the probability of winning. The rule of thumb is to be one of four or fewer serious competitors. But if expected profits are exceptionally high, or bidding costs exceptionally low (as with standard buildings), participation as one of about a dozen is often sanctioned. Being one of twenty is considered insane. When bidding departments experience slack periods that are known to be temporary, that is, with specifications for a large project due in a few months, then the department may be kept busy in the interim on projects with submarginal expected profits or with low odds.

Thus bidding departments are themselves part of overhead costs, variable in the long run (over a year) but not in the short run. In the long run contracts won must cover bidding costs. Even without competitive bidding, firms have to spend money on presentations to become prequalified and shortlisted.

Assessment of the competition will often lead a firm to reduce markups. According to a World Bank official, prices in 1986 were 20–30 per cent below those of 1984. They had expected to lend $400 million for an Asian dam, but the low bid came in at $290 million. The markups, which cover both overhead and profits, are never set by the bidding department but by top management in confidential meetings. Afterwards the markups are related to workhours via simple multipliers, but more sophisticated computer programs are being considered.

Business strategy for getting benefits from national political and financial institutions has not yet been discussed. Those topics will be left to the remaining sections of this chapter.

4 Government Support

Construction contractors throughout the world tend to see government as an agency for promoting their interests − opening and protecting markets, facilitating currency transfers, arranging insurance, and extending loans. They are mercantilists. The contributors to this book have reported that builders from all countries always consider their own government seriously remiss compared with others in

promoting construction interests abroad. The international departments of trade associations would, of course, be out of the lobbying
business if they said anything else; but their constituents make the
same accusation with vehement sincerity. Obviously, not all governments could be below average as construction promoters. In this section we shall compare basic official attitudes and resulting policies for
trade promotion, tax reduction, and insurance. The most critical
policy differences on finance and research are left to the next sections.

Attitudes and Basic Policies

For some governments the construction industry is just another mili-
tant special interest group that needs to be mollified now and then.
For others, the industry and the landmarks it sets up are a source of
national pride, worthy of special attention. Some officials un-
doubtedly think of it soberly as a foreign-exchange earner and the
chief instrument for capital formation. Most need to learn that it
operates under conditions so unique that policies suitable for promoting other sectors will not fit.

In general, governments of latecomer countries are more interventionist than those of advanced countries unless these are trying to raise
their international market share. In this vein, the French and Japanese
governments lead their contractors abroad in a forceful way.
Although Colombard-Prout reports in his chapter that French support was being reappraised in 1987, according to Seymour, that
government support was so extensive, both direct and indirect, 'that
French contractors in the past have been likened to an exported
nationalized industry' (Seymour, 1986, pp. 166, 178). Government
agencies, consultants, contractors, material suppliers, and bankers all
worked together. The Italians achieved similar results by a different
route. Italian contractors − including the nationally-owned conglomerates − had to push their government into providing support,
a process that did not gather momentum until the late 1970s.

Japan was more like France. The construction recession of the
mid-1960s led first to a thorough examination of the industry's problems and then to a coordinated government-industry effort to begin
overseas construction. MITI (the Ministry of International Trade and
Industry), the Ministry of Finance, and the Ministry of Construction
launched an Overseas Construction Promotion Policy co-ordinating
tax exemptions, subsidized credit, insurance support, and targeted
foreign aid. Co-ordination was promoted among contractors,
designers, equipment makers, and materials suppliers since the
Japanese defined the industry broadly. The government sought to
keep a low profile abroad by working through the Sogo Shosha
private trading companies and by encouraging contractors to begin

with joint ventures abroad. Nevertheless, the campaign for capturing overseas construction markets was launched in a spirit of a shy insular people boldly challenging great odds, of taking on formidable opponents – the low-cost Koreans and the entrenched but wily Westerners – all in politically risky terrain. In the late 1980s with a more serious construction recession at home and an unfavourably strong exchange rate, the Japanese formulated a revised long-run overseas construction strategy looking toward the twenty-first century.

The absence of a Ministry of Construction is usually a good indicator of lack of interest in promoting the sector. The British, German, and Scandinavian governments have not given high priority to their international contractors, and the American government has been least supportive, with the Reagan administration for a time even seeking to withdraw what little help there was. The policy consequences for all these countries will be reviewed below.

Most interventionist have been the Koreans, with detailed government permits and controls that establish contacts, entry, exit, prices, and wages throughout the world for their companies. As Chang puts it, the Korean government 'makes its presence felt in every aspect of overseas construction work'. The Turks resemble the Koreans with such government-sponsored export promotion. Other developing countries, like Brazil and Tunisia, give top priority to capturing and keeping their own home markets; but among these, the larger countries with their opportunities for specialization and economies of scale have more scope for displacing foreigners.

Trade Promotion

Least controversial in international competition is the promotion of trade by gathering and spreading information. Since knowledge is most productive for society if widely available, centralized gathering and distribution of information are common. Trade associations and cardinal periodicals (like *ENR*, *International Construction Week*, *Construction News*, *World Construction*, *Le Moniteur des Travaux Publics et du Bâtiment*, *Baumarkt*, *Bauwirtschaft* etc.) can do much of the job, but for quick targeted action, they are no substitute for well-trained commercial service officers at embassies. Contractors are unanimous about the importance of having information about possible projects during the gleam-in-the-eye stage, long before bids are formally requested. The British foreign service and the two dozen specialized construction attaches at Japanese embassies are said to be especially effective at ferreting out construction plans. A better foreign commercial service has been the main request of the Export Committee of the German Construction Industry Association, while the Americans under the Republicans had to lobby hard to keep the Foreign Commercial Service from being dissolved.

The contrast between interventionist and reticent government–business co-operation can be seen in Japanese and American missions for trade promotion. The Japanese missions are more frequent, larger, and have a co-ordinated follow-up based on the information collected. American firms participating in missions consider their findings such 'proprietary information' that obtaining useful follow-up reports from them was considered futile by US government officials.

When major projects are government sponsored, as can be the case for 80 per cent or more of the work in developing countries (World Bank, 1984, p. 39), then diplomatic pressure for information easily turns into pressure for the contract itself. Partly to facilitate government-to-government procurement of construction services, the Koreans increased the number of their Middle Eastern embassies from 6 in 1975 to 15 by 1981. The construction industry is well represented in the Korea–Saudi Co-operation Committee for Economic Development and in a number of other agencies, foundations, and institutes.

Other countries may send a high-level official to clinch a deal with a pre-emptive offer before competitive bidding. Such an attempt was made unsuccessfully by the British before a Japanese–Italian consortium won the contract for the second Bosporus Bridge in 1985. In this case, the Duke of Kent was sent to Turkey as head of a trade delegation pleading the case for Trafalgar House, PLC. As the *Wall Street Journal* reported, 'the presence of the tall, debonair duke, a first cousin of the Queen, opens doors around the world' (29 May 1985).

Constructors and their lobbyists also believe that governments should press institutions like the World Bank to make sure that contracts are awarded in proportion to national capital contributions. Americans, Canadians, Italians, and others have liaison officials attached to their bank directors' offices to get early information, to monitor tendering procedures, and to press contractors' complaints. Contracts do not necessarily go to the lowest bid, but the 'lowest evaluated bid', which includes completion time and other factors. World Bank procurement officials, however, insist that no preferences are given and that national laws to the contrary are superseded by international treaties.

Measures on Counter Trade, Collusion, and Corruption
Governments are effective in promoting their overseas contractors in connection with bilateral trade agreements, especially the acceptance of commodities as payment, called 'counter trade'. Until the mid-1970s barter deals were accepted by few firms and with only limited success (Neo, 1975, p. 239). Such deals have consistently been used by non-market economies such as the Soviet Union, for example with Finn Stroi and with the Italians at the Voljski steel works. For

the past decade, however, governments have helped to negotiate them, especially in the developing world. Norsa has reported that the Italian state-run conglomerates, ENI and IRI, accept counter trade payments under conditions that the private sector would consider unprofitable. Italy also has an official programme for helping private construction firms negotiate counter trade deals, as do France, Japan, Korea, and Brazil. The West German and British governments do not deal in counter trade commodities but have an advisory service for bringing contractors together with private commodity brokers. The United States lets them figure it out for themselves.

As might be expected, unlike others, the West German, British, and American governments are reluctant to promote trade by encouraging national consortia of overseas contractors, with the Germans doing the least in this case. The British foster some consortia as part of certain financial aid packages. The American Export Trading Company Act of 1982 grants immunity from antitrust suits to exporting joint ventures, but so far no American construction firms have taken advantage of this law, except as a preliminary step toward the possible Three Gorges Dam project in China proposed by the Bureau of Reclamation of the Department of the Interior. The British are the only ones who do not have a public or quasi-governmental consulting firm, such as the Bureau of Reclamation, the US Army Corps of Engineers, or the Naval Facilities Engineering Command.

An American peculiarity is legislation that actually discourages trade if boycotts and bribery are required. American firms are not allowed to comply with bans against dealing with Israel, and heavy criminal penalties could go to company presidents caught paying bribes to foreign officials, directly or via any chain of intermediaries. Japan is the only other country with such legislation although in substantially milder form. In most countries paying bribes for an invitation to bid and later to be selected is condoned practice. In an extreme case, a British firm agreed to pay a $112.5 million commission to a Middle Eastern planning minister to be allowed to tender for a $2.5 billion railway project. When the company refused to pay a further eighty million dollars to another governmental faction, the deal collapsed, and the project manager was given a 20-year prison sentence for 'economic espionage'. Indians subsequently designed the railway, and Brazilians built it (Aburish, 1985, pp. 105–19; *The Guardian*, 27 October 1985). Most tolerant of bribes are the Italians, who consider such payments, if necessary, a legitimate deduction from taxable corporate income (Demacopoulos and Moavenzadeh, 1985, p. 177; *ENR*, 10 July 1980).

Tax Policy
Better than opportunities for making money through trade promotion

is receiving money directly. It is less uncertain and less trouble. Governments can dispense money most easily by just collecting less in taxes. Lower taxes on salaries abroad lets firms reduce bids and win contracts.

Taxes and tax exemptions come in many forms, and most cannot be discussed here. One form of tax is forcing companies to convert foreign exchange to the home currency at unfavourable exchange rates, as Brazil does (while not otherwise taxing construction export earnings). But usually taxes are more straightforward than tax exemptions. Italy and Japan do not tax income earned abroad and have further tax concessions on the domestic income of exporting companies.

Most reluctant to reduce taxes is the United States. It has the least concessions on corporate income taxes abroad. In 1982 American corporate income taxes on construction exports came to 0.4 per cent of turnover (Price Waterhouse, 1985, p. 16). The United States also taxes personal income abroad, including allowances for housing and schooling. Despite an exclusion of $70,000 per worker, this tax is bitterly resented by the executives whom it hurts most. Nevertheless with American wages and fringe benefits abroad averaging only 4–5 per cent of turnover (and much less in competitive settings, due to employment of Europeans), it appears doubtful that these taxes cause Americans to price themselves out of the market.

What makes tax exemption effective is applying it in such a way as to induce modernization. The Koreans, for example, have a 2 per cent of turnover exemption, provided funds go into a reserve for modernization and promotion. Similarly the Japanese can defer taxes for 5 years on income shifted to a contingency fund against losses abroad. Moreover, 20 per cent of income may be deducted as a fund for expanded research and development. In effect, all overseas corporate income can go untaxed. But apparently competing countries are not impressed by these tax privileges; as an example, the German overseas contractors do not press their government for comparable deals.

Insurance and Bonds
The riskiness of construction for both builder and client must be a central theme of any policies aimed at promotion of the sector. Risks too large for one person, firm, or project can be reduced through pooling or collective action. The uncertainties and amounts involved in international construction projects are enormous, however, beyond the capacity of private insurance companies. Governments of every major exporting country therefore provide both credit and political risk insurance. But insurance is not an arena of intense international competition. Indeed, such competition is formally restrained through the *Union d'Assureurs des Credits Internationaux*, known as the Bern Union.

Most insurance comes from the institutions that have given export credits and covers only some part (mainly 70–95 per cent) of such loans. The insurance agency may be public or, as in the United States, a consortium of dozens of private insurers – the Foreign Credit Insurance Association – whose loans are reinsured by a public institution (the Export–Import Bank of the United States). Both the Ex–Im Bank and the Overseas Private Investment Corporation are American government agencies that insure against political risk. Unlike most major construction exporters, the United States provides no official bond or exchange risk insurance.

British exporters are covered by the Constructional Works Policy insurance of the Export Credit Guarantee Department (ECGD). Up to 90–95 per cent of losses are covered if the client refuses to pay, becomes insolvent, is forbidden to pay in foreign exchange, cannot pay because of war or other disturbances, or because of cancellation of various necessary permits. Insurance rates are in proportion to the creditworthiness and political stability of countries as judged by the ECGD advisory council.

French insurance coverage is most comprehensive, including inflation risk insurance and uniquely, no ceiling on amounts insurable. The Italian and Korean coverage is lowest among the major countries included here although other countries like Canada, Switzerland, and the Netherlands also have limited coverage (Demacopoulos and Moavenzadeh, 1985, pp. 145–168). Yet dissatisfaction does not seem to be intense.

Only the Japanese and Koreans have insurance against the cost of lost bids. Nevertheless, no one complained about inadequate insurance coverage as much as the Japanese, according to our contributors, Hippo and Tamura, who concluded their chapter on this note. MITI provides insurance against exchange losses, bond risks, defaulting clients, and losses or damage due to wars and other causes. But the value of this insurance is doubted by the Japanese firms who have it, a minority, because of MITI's narrow interpretation of contingency clauses. For example, the Iran–Iraq war was classified as a mere 'dispute', hence no ground for reimbursing contractors for interrupted work. No doubt that interpretation played a major part in persuading Japanese firms to reduce further commitments in the inflammatory Middle East.

5 Finance

Since the early 1980s the most critical short-run element throwing contracts here or there has been the provision of low-cost finance.

Established contractors would, of course, rather compete on the basis of quality and their network of special contacts. But as conditions tighten from a seller's to a buyer's market, first price becomes an increasingly important issue of contract – hence worldwide procurement of materials and low-cost but disciplined labour – and then finance. All contributors to this book have reported that contractors now feel that the most essential government measures are ones easing terms of credit. As mentioned before, for the client a lower price or cheaper finance are interchangeable ways of reducing monthly debt service payments, provided finance is available at all.

Competition with financial contacts is neither new nor unobserved. For centuries builders have introduced their clients to bankers who could let plans become reality. The builder who comes up with the financier naturally has the inside track for the job. In the mid-1970s, Neo observed that clients outside of the oil-producing countries frequently asked contractors for help with finance, and 56 per cent of international contractors in his sample of 22 had done so (Neo, 1975, pp. 112, 116.) Ten years later Seymour counted 18, out of his sample of 20 contractors, calling finance their main problem. He concluded that 'contractors are competing on the conditions of the financial package they offer more than any other aspect' (Seymour, 1986, pp. 219, 235.). Half of the top 250 international contractors surveyed by *ENR* said that in 1983 they had to arrange finance, and several called this effort at least as important as selling their technical skill. An executive of the British firm, Davy, said, 'We've always taken the view that financing a contract is integral with selling it . . . If you're on the bid list at all, your technology is acceptable. Therefore it really comes down to price and finance' (*ENR*, 2 August 1984, pp. 30–2).

Role of the Capital Market
Construction more than other sectors involves finance because of the greater lag between production and sale. Expenses begin as soon as plans are drawn, the site is prepared, materials are ordered, and foundations are being dug. The project may not be complete for two or three years, even four or five. That is the construction lag. After completion, a project will yield its output for decades, perhaps centuries, causing an income lag. Working capital is needed to cover the construction lag; project finance for the income lag.

Rich clients with outstanding projects, such as OPEC countries during the late 1970s, have provided 20 per cent cash advances for working capital, timely progress payments, and cash when the project was complete. If the project looks highly profitable, but the client lacks funds; the private capital market is there to earn commissions channelling working capital and project finance in that direction.

American, Japanese, and British contractors are outstanding in using their finance departments and banking contacts to give clients access to market funds at lower rates than they could otherwise obtain. But every country does some of that. As early as the 1920s, Drewer reports that the Axel Johnson Group involved the Swedish Enskilda Bank in a Rumanian project. The client benefits from lower interest rates and longer maturities, not because of an implicit subsidy, but because the market flow of capital has been improved. Large contractors from capital-surplus countries benefit the most because they are most trusted by well-heeled financial institutions.

In all cases it is the creditworthiness of the client and the project, not that of the contractor, that matters the most. The project must generate an income stream with a present value equal to or higher than the loan. Neither banks nor rich and creditworthy contractors lend for hopeless projects and clients just to keep building activities going.

Bilateral Loans and Aid

Essentially, not lack of finance, but lack of viable projects stifles construction demand around the world. Financial institutions, apart from the soft-loan windows of development banks, will not lend for ventures that are too uncertain, too long-term, too round-about, or too big compared with yields. If projects are to be built anyway, governments have to assume the risk and make up the difference between construction costs and the present value of the expected income (repayment) stream. One way of doing that is to reduce debt service payments to a rate that project earnings can actually cover. Part of the amount owed can be forgiven as a grant, or interest charged can be put below the market rate. The effect is the same: a gift.

Any national government that collects enough in taxes should be able to keep a high volume of construction going in this manner. The national treasury simply has to replenish the capital of lenders, public or private, now and then. Problems arise if the programme has a multiplier effect on national income, inducing imports; or if foreign contractors have to be paid in foreign exchange. In that case, exports have to rise; or funds have to be borrowed abroad, raising the question of foreign debt service capacity. This is where foreign government-supported financial institutions come in to get the projects going and find work for international contractors. The fellow citizens of the international contractors now pay the taxes, reducing consumption to make a net outflow of resources possible.

As described in the country chapters, official financial support mainly consists of a mix of insurance measures (already discussed in section 4), medium-term export credits, and foreign aid. Subsidized export credits are provided primarily for manufactured goods, but virtually every country has extended the system to construction services.

The OECD Arrangement

Pressure on government institutions for ever lower interest rates became so intense that OECD countries decided to soften this form of competition in 1978 with an 'Arrangement on Guidelines for Officially Supported Export Credits'. Minimum interest rates between 9.85 and 12.25 per cent and maximum repayment periods between five and ten years were established for borrowing countries in three per capita income categories.

These uniform rates did not allow for variations in inflation and corresponding national market interest rates, so the 'arrangement' was modified accordingly in 1983 and again in 1987. Interest rates for loans in any currency are now tied to a commercial interest reference rate or to a weighted average of medium term government bond yields. The French and Italians prefer to denominate their loans in whatever foreign currency permits a lower interest rate.

The scope of this agreement was limited because only loans with a grant element of less than 25 per cent were covered by the arrangement. If interest rates were below the approved rate, the present value of the stream of repayments was calculated; and if that came to less than 75 per cent of the loan, the grant element was assumed to exceed 25 per cent. In that case any low interest rate could be charged. In March 1987 OECD countries agreed to extend coverage to loans with grant elements up to 35 per cent.

Competition with Subsidized Finance

The French, followed by the Italians and Japanese, are most zealous in fostering exports of construction services with subsidized finance. Least supportive among major countries are the Americans. The French have the most comprehensive export credit and insurance system, and they combine that with generous amounts of foreign aid. They are the inventors of mixed credits, and their share of untied aid has remained relatively constant at only 40 per cent. The comparable US share in 1983 was 56 per cent (up from 26 per cent in 1975), and that of the Germans and Japanese a surprisingly high 70 per cent. However, the Japanese subsidize feasibility studies with an annual $200 million, compared with $100 million by the French and a mere $18 million by the Americans (Jarboe, 1987; OECD, annual reviews of Development Co-operation).

From time to time, one reads of odd arrangements. For example, the Japanese called a five-year 6.5 per cent credit to China 'import financing' on the grounds that the output from the project would eventually be exported to Japan (Pearce, 1980, cited by Seymour, 1986, pp. 246, 249). In the case of the Bosporus Bridge in 1985, the competing Japanese and British bids were only 4 per cent apart at something like $115 million. For the 130 miles of access highway, however, the Japanese–Italian consortium was prepared to charge $50 million less than others or $440 million. More important than price was the $205 million Japanese loan at 5 per cent, repayable over 25 years with a seven-year grace period. Through Impregilo the Italians provided another $130 million at heavily subsidized rates. Like the French, the Japanese simply call contributions a legitimate form of foreign aid, not a subsidy (*Wall Street Journal*, 29 May 1985). They further preserve their low profile by letting the Italians consider the project mainly theirs, as may be seen in Norsa's contribution.

Italian financial support for overseas contractors did not become strong until the Ossola law was supplemented in 1981. Now their pre-bid expenditures are heavily supported, and most risks are insured. Government credit is tailored to specific projects, primarily with subsidies that bring interest charged by co-operating lenders down to competitive levels. More than other countries, Italian contractors will therefore assemble credit packages and bid low in the anticipation of later government support. With this strategy, Italians were able to preserve their Latin American and African markets, while German fortunes slipped with those of their oil-exporting clients.

Hesitant Subsidizers

The German government, like the British, American, Swedish, or Danish has not been aggressive in promoting exports of construction services with subsidized finance. German concentration on Saudi Arabia and Nigeria kept finance from being an issue during the boom years. Success turned to lost business with the decline of oil and the Iraq–Iran war. Only in the case of delayed payments and stalled dam construction because of that war did the German government intervene with finance to reschedule a major debt. Otherwise, the Germans took no special measures in the face of declining construction exports after 1982, according to our contributors, Riedel and Gluch.

The British and Americans not only subsidize their construction exports less than others, but what strategy exists is less co-ordinated. The British Department of Trade and Industry sees tied aid as a legitimate way of promoting trade, along French and Italian lines, but the Treasury views the Aid and Trade Provision (ATP) and tied aid as dubious. The ATP amounts to some 5 per cent of British foreign aid and, like the US Exim Bank's war chest, was set up in 1977 for matching 'predatory financial terms' offered by the French. In 1987 the British finally signed a cofinancing framework agreement with the World Bank, and the US Exim bank consolidated and extended its programmes. Both the US and the UK, however, devote a smaller share of their national products to foreign aid than others, and their preference is more for raising the health, education, and productivity of poor people than for pushing the fortunes of exporters. Such a policy may be noble but it is not politic. Speaking of the mixed-credit programme, a French Embassy official said in Washington, 'It is difficult for us to understand why such a violent campaign has been launched to reduce this form of aid' (*New York Times*, 2 February 1986).

Needless to say, the developing countries or NICs represented in this book, plus others not included such as Turkey and India, are least able to promote construction exports with government finance. Indeed, Turkey has no export-credit system at all. When Libya delayed payments to Turkish contractors working on $11 billion worth of construction in 1982, everything came to a halt until the Turkish government worked out a counter trade arrangement involving oil (*Middle East Economic Digest*, 1 October 1982). Brazilian overseas contractors are supported with cruzado financing, but according to Verillo, the amounts involved are too uncertain and modest to spell the difference between contracts won and lost.

The Koreans are also handicapped by capital scarcity. Export credits and insurance are available but no loan guarantees or insurance against the vagaries of inflation and exchange rate adjustments. Korean contractors had to raise their bids when 20 per cent cash advances were no longer available from Arab clients. Indeed a quarter of the debts owed to them were in arrears. According to Chang, the government had to compel Korean banks to reduce interest rates substantially below rates available to Korean firms abroad. These banks therefore pooled their resources to provide new guarantees, backed up by the Overseas Construction Promotion Fund. To make constructors more creditworthy, the Korean government pressed insolvent ones to sell off their assets, to merge with larger firms, or to let banks take them over.

International Development Banks
One set of institutions combines the traits of capital markets with those of public development assistance. These are the international development banks and funds. Most important is the World Bank and its soft-loan *alter ego*, the International Development Association (IDA). Next come the regional development banks for Africa and Asia and the Inter-American Development Bank (IDB). West Africa, East Africa, the Caribbean, and Central America have subregional banks. The Japanese, OPEC, Saudi Arabia, Kuwait, Abu Dhabi, and others have also set up funds and banks for development. But the World Bank and Regional Banks do the bulk of the business, around three-quarters.

For its annual lending of some $16 billion, the World Bank mainly uses repayments of past loans and proceeds from bonds sold in capital markets. Projects must therefore earn enough to cover amortization and interest, preferably without harmful effect on a country's international debt servicing capacity. However, over a fifth of World Bank loans are made through IDA to 50 of the poorest countries. These loans charge no interest, have a ten-year grace period, and are repayable over 50 years. Obviously they have a large grant element, and the capital of IDA must be replenished periodically if lending is to continue.

Normally the Bank contributes between a quarter and a third of project finance, with the rest coming from local sources or other donors, often in the form of cofinance, either 'joint' or 'parallel'. If financing is joint, a common list of goods and services is procured through procedures permitted by the Bank's Procurement Guidelines usually international competitive bidding. If financing is parallel, however, the guidelines do not apply to the designated non-bank part of a project, and loans for that can be tied to the donor's exports. In the past decade, cofinancing was involved in hundreds of projects (not all in construction), worth about $28 billion, often under the Cofinancing Framework Agreements signed with nearly a dozen countries. The leading co-financers were the Italians with some $450 million in loans during 1982–6, according to our chapter on Italy by Aldo Norsa.

When approving contract awards in accordance with its Procurement Guidelines, the Bank does not look behind bids for subsidies or deals. In one case a consortium with local African participation lost to a French bidder by a mere $2000 on a $3 million project. The Bank required acceptance of the lowest evaluated bid by the French contractor as the only way of maintaining credibility with that important constituency, the industrialized countries. Bank executives said unofficially that none but construction firms ever made a political fuss about lost procurement awards.

The importance of World Bank loans for international contractors is less than one might expect. Much of its lending is non-project or goes for education, health, small-scale enterprises, and other purposes with little impact on construction. Of the loans for civil works about three-quarters are executed by host-country firms. Small projects are awarded under local procurement procedures if no foreign interest is expected, but nevertheless foreign firms can bid. Many large projects are 'sliced and packaged' to permit both small and large firms to bid. There is a 7.5 per cent bid preference for local firms of countries with a per capita product below $400. What remained to be disbursed to international contractors in 1985 was $670 million ($600 million in

1982 dollars). If *ENR* data are adjusted for double counting and for the lag between awards and disbursement, international construction turnover in 1985 was about $45 billion. Hence the share of World Bank civil works disbursement was no more than 1.3 per cent of receipts by international contractors. Even as a share of Latin American, African, and Asian International construction (without the Middle East), the direct share of the World Bank is only about 3 per cent. Including funds from host governments and co-financers, value of the projects as a whole probably rises to somewhat over ten per cent of international contracting turnover.

Awards of World Bank construction contracts went faithfully to the lowest evaluated bid, but the process was not designed to be immune to the bold and resourceful tactics of firms from countries with cofinancing, interventionist, subsidizing governments. Although the Bank's process for evaluating bids should not be confused with the cofinancer's procedures, in the words of the Bank's *Operational Manual*, cofinancing was welcomed because it would 'increase the flow of capital to developing countries and . . . permits the dissemination of the Bank's own funds over a significantly larger number of projects and promotes more efficient use of total available external resources.'

From the early to the mid-1980s, the share of France, Italy, and Japan in civil works disbursements by the World Bank to international contractors rose from a third to over half. Their combined growth rate was an annual 17.3 per cent (current dollars), compared with 7.5 per cent for total disbursements to all international contractors. As can be seen in Table 11.2 the combined French-Italian-Japanese volume rose from about three times to eight times that of American firms. In constant dollars, the American share of World Bank civil works disbursement declined by 42 per cent during this short period. The last column of Table 11.2 shows how the trend continued in 1986.

Resource Transfers and Efficiency

In all this discussion of banks, export credits, interest rates, insurance, guarantees, tied aid, and national rivalries, one can lose sight of basic factors. A country can export more construction services than other countries because its construction services are in greater demand and able to command a higher price, or in greater supply, meaning offered at a lower price. In both cases, differences in quality, real or perceived, will play a part. In the long run those factors will prevail. In the short run, however, some countries are willing to transfer resources abroad because they are great savers or aid-giving tax payers. They can charge a lower price for finance of any export, including construction services. They are willing to work more and consume less. That makes them competitive suppliers. Japan and France fit that role.

Table 11.2 *World Bank Disbursements for Civil Works:*
Payments to Contractors by Nationality
($ Millions per year)

Country	1980–81	1984–5	1986
France	64.1	115.5	110.2
Italy	58.3	106.6	110.9
Japan	38.9	83.6	131.2
United States	55.8	38.2	28.5
United Kingdom	19.0	27.4	18.4
Spain	—	15.7	5.6
Canada	18.2	11.7	7.3
West Germany	20.6	9.0	26.9
Belgium	7.8	8.6	16.5
Netherlands	18.5	4.2	5.7
Switzerland	9.8	4.0	5.5
Sweden	12.9	0.5	0.9

Source: World Bank, Data Analysis Unit, Loan Department. Years are fiscal.

Japan leads the world in saving and is therefore the leading exporter of capital; France gave 0.77 per cent of its GNP as official development assistance in 1984, compared with only 0.23 per cent for the United States.

Countries that save less and operate their government at deficits, like the United States in the 1980s, are resource absorbers, capital importers, not exporters. All exports will be harder for such countries. Those goods and services that retain and expand their foreign markets despite ensuing high costs must benefit from demand for special abilities and qualities. In construction, as in many other sectors, these abilities and qualities are associated with technological change which, in turn, depends on research and development.

6 Technology

Construction technology is productive knowledge about the use of labour, materials, and equipment to make buildings and civil works. Since knowledge production and diffusion is sub-optimal if limited to what is profitable in free markets, governments usually get involved. They run schools, do research themselves, pay for research, or grant patents that allow cost recovery for research through monopoly pricing. In the short run, government support with finance is more popular with the industry because it means cash in hand. But unlike the more uncertain but steady growth of technology, subsidies cannot keep growing since their share of national product is bound to have a ceiling.

Continental European Research Support

The French have raised their share of the international construction market, not only because of greater financial support, but also because of greater technological support. According to one comparative study, 'the systematic use of science in support of economic development has been carried further and has been more rationally organized in France than anywhere else' (Solo, 1975, p. 103). In other continental European countries, research is not likely to be carried out by the government directly but through combinations of autonomous institutes, university laboratories, and private firms that receive public matching funds for their work. The Germans, for example, heavily support research in this fashion but have no major government construction research laboratory. The work, nevertheless, tends to follow priorities set in a 1981 report, 'The Future Tasks in Construction Research', commissioned by the Ministry for Technology and Research. Development of products and processes that can be patented is encouraged since that helps in international competition. The Germans, partly as a result, had the highest rate – 80 per cent – of winning major contracts after being the designer-consultants at an earlier stage. German design encourages German patented products and processes to be used in projects built by German contractors. The Japanese linkage was 63 per cent, the French and Italian 50 per cent, the American 43 per cent, and the British only 13 per cent (Murphy, 1983, p. 138).

Results of various European research efforts include superior tunnelling methods, large robotized cranes, pumps with great lifting capacity, slide formers, slip forms, novel compacting devices, better prestressed concrete technology, and novel casting methods for computer-controlled variety.

The Japanese Technological Strategy

The Japanese modernized their construction industry by first importing American technology and then moving on to European methods when these proved superior. Now something like 3 per cent of construction turnover is devoted to research – compared with 0.4 per cent for the US (Tucker, 1985, p. 4). There are no government research laboratories or matching grants for private organizations, but firms can deduct from taxes a quarter or half of research expenditures if these go beyond a company's peak during the past decade. As a result some firms have research laboratories employing hundreds and spending millions of dollars annually.

Research priorities have been specified in a five-year plan of the Ministry of Construction for advancing technology, but executives of major Japanese firms have told us that these priorities are not mandatory for individual firms. In their laboratories and on the production site, firms study core technologies and applications, often jointly with manufacturers. Themes are information processing, modularization, automation, and robotization. The Japanese are currently perfecting automated earthmoving equipment utilizing sensors, shotcrete robots for tunnel spraying, slab finishing robots, reinforcement placing robots, robotic underground excavators, undersea robots, and robotic tile inspectors. They are also seeking to systematize design methods, to integrate cost management with the construction process, and to anticipate future maintenance problems at the design stage. Some of these design and management goals involve data-intensive activities in which Americans still have the lead. But the Japanese do not expect the sort of quick payoff from research that seems mandatory to Americans (Halpin, 1986b, pp. 6–11).

One must not, however, exaggerate the role of technology. During 1985–7, Kajima, for example, had a negotiated lump-sum design-construct contract for a $500 million automobile plant for Mazda south of Detroit, Michigan. According to Kajima officials interviewed at the site, no special Japanese construction technology was being used – nothing in setup, process, equipment, or materials – not one of Kajima's 1,559 patents. All site work was executed through some hundred American subcontractors operating in their usual fashion. More authority was delegated to them and the trade unions for managing men and materials than would have been delegated in Japan. As a result, Kajima had a staff of only 64 (14 Japanese) on the site instead of the 130–150 that they might have employed in Japan. The result was loss of quality and a more uneven (but not slower) schedule. Costs would have been equal to those in Japan with a 180–200 yen per dollar exchange rate, but lower at the 1987 rate of 150 yen per dollar and higher at the 1984 rate of 250 per dollar. In terms of dollars, costs were not less than those of an American competitor like Austin. Kajima's main advantage was smoother communication with its Japanese client, for whom they had designed and built the Hofu plant in Japan together with Ohbayashi and Shimizu, and with Japanese equipment vendors, for whom they had a separate construction management contract. The point is, they had not won the contract because of superior construction technology.

American Technology and Organization
American construction and design firms do some one-off problem solving but virtually do not carry out genuine research that raises

productivity on the site. Only the Bechtel Group and M. W. Kellogg have made substantial commitments to fundamental research; but much of that is for process, not installation or construction. Of the universities, only the Massachusetts Institute of Technology has a long major tradition of trying to involve building firms closely in engineering research. Of course, much consulting by professors goes on elsewhere but is viewed as a diversion from the academic mission. Moveover, firms complain that the academics lack skill at estimating the risks and payoffs from their ideas. Government research has been limited to military structures, public utilities, testing, and establishing standards. Lately the Department of Defense, the National Science Foundation, and an industry-client group in Texas have set up research centres at various universities, but the momentum behind these efforts remains in doubt (Jarboe, 1987).

Insofar as American firms retain a technological lead, it is due to equipment makers, material suppliers (chemical additives, industrial textiles), computerized management efficiency, and familiarity with complex, high technology in the design of structures. Integrated American turnkey builders are helped by being ahead of the Japanese and Europeans in computer aided design and drafting (CADD). Better 'constructability' due to CADD and design-build integration, however, leads to adaptations that, while important, are not research leading to diffusion of innovations. The fast-track schedules of industrial projects already mean working with a changing design on the site, and the procedure is complicated enough without further experiments.

In any case, design and installation of mechanical and electrical equipment, especially pipes and cables, dominate process plant construction, not the creation of space within a structure. Ownership of process technology is what wins design contracts for firms like M. W. Kellogg, Badger, Stone and Webster, Lummus Crest, and Foster Wheeler; and the construction contract (worth ten times or so as much) is tied in so that less pressure exists for further innovative cost reduction. One should recall that in 1985 over half of American overseas contract awards still consisted of building power and process plants and project or construction management.

Although shared with the British and Canadians, the American lead in management based on systematic analysis of data has not so far been challenged by others. Halpin has suggested that the US advantage in this field is due to the training of engineering professionals who are competent computer scientists and managers (Halpin, 1986a, p. 11). In other countries these skills are often in different, competing, possibly even antagonistic groups. For management purposes the computers and software must be complemented by data bases to facilitate purchasing, component tracking, expediting, quality control, warehousing, document control, accounting, cost control, field engineering, and personnel. Since it is not uncommon for most site labour to be wasted due to scheduling delays, the yield from this skill has been great; but its complexity makes it one not easily transferred. To catch up with American skills in this area has been among the motives for acquiring American firms on the part of European constructors.

Britain: A Similar Case
The British situation has been summarized in 'Long-term Research and Development Requirements in Civil Engineering', a report prepared in the mid-1980s (undated) for the Science and Engineering Research Council and the Departments of Environment and Transport. Annual research and development expenditures were estimated at £150 million, with about two-thirds coming from private materials and components makers. In general, the situation is like the American, except for the important contribution of the government Building Research Establishment, especially the Research Station at Watford. Otherwise Britain has a similar lack of co-ordinated support at the national level. Contractors are not research-oriented, but the UK shares the lead in computerized management systems with the US. The British recognize that the 'growing capability of developing countries will mean that consultants, contractors and material producers will have to go "upmarket" to compete successfully in international markets'. Design-build has made the least inroads into British practice compared with other countries, however, and British consultants actually pride themselves on designing structures within the technological capability, hence bidding range, of builders from many countries. Hence their low 13 per cent design-construct linkage, mentioned above. A reputation for being unbiased yet cheaper accounts for the high British share of international design billings – 12.7 per cent – twice the German one.

Export Promotion and Technology
The true national benefit and function of the controversial export subsidies of France, Italy, and Japan may well have been their indirect promotion of technological change. Otherwise these subsidizing countries – in accordance with well-known free-trade rationale – may have captured larger market shares at excessive cost. The Japanese MITI certainly had a policy of identifying technological sunrise industries and promoting them with trade policies so that Japanese firms could gain market shares and experience, sliding down their 'learning curves' with innovations.

Other countries, however, will not necessarily catch up best with these three leaders in on-site technology by copying their market-snatching strategies. Such policies may benefit contracting firms but not nations as a whole. If innovations, technological change, and learning are really needed, they can be sought better with policies that encourage them directly or that remove obstacles that stand in their way. Trade subsidies might have been sufficient to induce further research in the Japanese and French setting, but in other countries they might prop up obsolete ways of doing things and let firms and governments postpone hard decisions for supporting technology.

Collective action looking far ahead and generously supported by contractors, designers, equipment makers, material producers, research institutes, and government may be what is needed (and more for construction at home than for technology transfer abroad, as we shall stress in the next section). Such integrated support of technological change might have existed all along in Japan and France. The problem for countries like the US and the UK may not be that of an unlevel playing field abroad but inertia, industrial fragmentation, anti-trust laws, recalcitrant labour, and timid management. Changing all that should be an objective of policy. Benefits are likely to follow regardless of whether export markets and international rankings rise or fall at the same time. Technological change often involves problems, but it is far more likely to raise the general welfare than trade-distorting measures.

Transfer of Technology

A reputation for having an advanced construction technology is an important ingredient of international competition for contracts, but such technology is not systematically developed with that in mind. Under the foreign aid programmes of France, Britain, and a few others, government laboratories have developed labour intensive, local-materials-using building techniques for the tropics; but these are mainly for housing and roads of little interest to international contractors. Apart from that, technology developed in every major country primarily reflects what is needed by its own domestic market. It reflects demand, the types of structures a country wants; and supply, the way construction and related industries have developed. Hence, the Dutch are best at coping with shorelines, the Austrians and Swiss with mountains, the Japanese with dense settlement, the Scandinavians with social facilities, and the Americans with high technology for merriment or armament. For every country, though not every firm, building abroad is secondary.

Technology that has been developed for the home market of advanced countries will often be of little interest to builders in poor

LDCs. Much of the Japanese aseismic high rise techniques, underground space development, automatic warehouses, and robotics are simply not relevant to low-wage, sparsely settled countries that still need infrastructure to go with a multitude of small and simple buildings, plus an occasional advanced production facility here and there. For such countries, learning the disciplined sitework efficiency of the Koreans or the scheduling skills of Americans is more important.

Virtually all international contractors claim to be staunch transferrers of technology, but that mainly means doing something familiar that seems strange in a foreign land, setting up unusual structures with freakish machines and gangs of alien workers. Local contractors are taught as little as possible about design, procurement, scheduling, and management of technology. After all, the internalization of technological change within a multinational enterprise is one of the main factors giving a competitive advantage (Buckley and Casson, 1976; Caves, 1982; Williamson 1981 and 1985).

Fostering technological transfer is now an obligation specified in international construction contracts, but results have been far less than what is possible. According to one recent observer, 'there have been many Arab projects in which it cannot be said that the "key has ever been turned over" despite ever greater payments in the form of management and other fees to foreign firms and consultants' (Nugent, 1986). In our Korean contribution, Chang reports that licensing or purchasing technology has not been an effective method of transfer. A report by Philip G. Abbott confirms that, 'By far the majority of transfer programmes take place within joint ventures' in which counterparts from different countries work together as equals as specified in a contract (Special Report 223, *Economist* Intelligence Unit, 1985).

Only working very closely with the foreigners, sharing responsibility, inculcates those skills that have to be learned by doing. Hence the Tunisian firms, as Ferchiou reports, are seeking to be included in joint venturers, not just as subcontractors. The Brazilians have prevailed at becoming joint venturers and consequently have learned enough technological tricks to dispense with the foreigners in one specialty after another. As one Brazilian analyst wrote, 'Associations with foreign companies . . . have brought a significant contribution to our technological development. The total or partial absorption of each technology, depending on the project, marks the end of these associations' (Sercovich, 1984, p. 594). No wonder advanced-country firms do more public relations than spirited follow-through about technology transfer. D. S. Greenwood of the British firm, Costain International, put it very succinctly: 'International contractors will transfer expertise and experience if that gives them the competitive edge or access to future work.' Otherwise they will not. (Comments at the Institution of Civil Engineers, London, 25 February 1987.)

7   Review and Outlook

Review of Findings

The business strategy of international construction contractors, we found, consists of entering foreign markets cautiously and supplementing special skills with missing components from joint venturers – be it political acceptability, finance, high technology, or cheap labour. Negotiated contracts or design-build are preferred, but the ubiquity of government clients or multilateral financing often makes competitive bidding unavoidable.

Governments in their support of overseas contracting firms, range from the resolute French and Japanese to the half-hearted American. The US has pressed the Japanese hard for access to the Kansai airport and other infrastructure contracts, but it lags in fostering research, reducing taxes, promoting trade, negotiating counter trade, providing insurance, subsidizing finance, and even co-ordinating what support there is.

Subsidized finance is what contractors want most of all, and its provision partly explains the high market shares of the Japanese, French, and Italians. International institutions such as the OECD and the World Bank have tamed 'predatory financing' somewhat, but cannot overcome the basic willingness of some nations to transfer resources abroad to keep construction firms employed.

Continental Europeans and the Japanese lead in developing new site-oriented construction technology, occasionally giving them an edge in penetrating the markets of other advanced nations. Many of these discoveries and inventions, however, lack relevance for developing countries. There, American superiority in construction management and process plant design has allowed continued dominance for many types of project. Despite innumerable and ingenious efforts to forestall competition, however, entry of new firms is enough of a threat to keep monopoly profits temporary phenomena here and there. After construction booms, bankruptcies and retrenchments have been common in every country. One cannot even be sure that the high profits of some cover the risks and losses of others.

Regardless of nationality and public relations claims, international contractors transfer as little technology as possible to potential competitors. But under pressure from host countries and joint venturers, secrets and tricks of the trade do leak out. The competitive edge is likely to be lost to local enterprises first in site execution of conventional structures and last in the initiation and organization of large-scale, complex, and novel projects. As national contractors gain experience, they become latecomers to the international market by entering neighbouring countries when a boom elsewhere diverts the attention of the international giants. Once a market has opened for them, some firms will eventually develop enough skill, trust, and reputation to be able to compete anywhere.

Self-Images of Enterprises
Contractors from the countries surveyed here see themselves and their situation quite differently from the description just presented, but the self-images of various countries are strangely similar. All see themselves proudly as having a special knack for building technology compared with other nationalities. All furthermore believe they have a special knack for dealing with workers, colleagues, and clients in the host country. Brazilians and Koreans think rapport comes from being fellow Third-Worlders; Scandinavians know themselves as admired paragons of social responsibility; Germans think their legendary reliability yields ultimate trust. Americans see gratitude everywhere for open no-nonsense practicality, and Italians believe they combine all the qualities just mentioned in the optimal way. Only the Japanese think they are unappreciated outsiders. Yet even they, like the others, see themselves as dedicated technicians, people with a heart who never lose sight of the human dimension.

By contrast each set of national contractors perceives foreign competitors as wily and ruthless, greedy for money, in cahoots with one another, and indifferent about quality. They see foreigners as acting with menacing co-ordination, while their own compatriots are regrettably individualistic, prone to compete too fiercely with their own kind.

Finally, all construction executives seem to regard their own governments as especially unhelpful compared with others. The British look on theirs as stingy and inconsistent; the Japanese feel cheated on that crucial element, insurance. Brazilians find their support haphazard; and Italians call government listless, inefficient, their primary handicap. Most extreme are the Americans who consider theirs not just incompetent but hostile. Nowhere did contractors doubt that their own prosperity would confer blessings on both home and host countries.

Uncertainties
What does the future hold? Will the industry continue to decline, recover, or perhaps again have a great boom? Attempts to answer

these questions must bear in mind that construction demand is deriva-
tive of the demand for the output from structures. Rising incomes,
population growth, and migration induce a greater demand for housing
services and urban public utilities. Growing demand for other goods
and services call for factories, warehouses, offices, and other corresponding capital structures.

In the mid-1980s the international construction market was divided
roughly in quarters – one for the Middle East, another for South and
East Asia, a third quarter for Latin America and Africa, and the last for
Europe and North America. That these proportions would not last was
certain. After all, the Middle East had already declined from nearly 40
to 22 per cent, or from $50 billion to $14 billion (1982 dollars). But
where would the next boom take place? China? Soviet Siberia? And
what would be its sectoral focus? Something tropical or polar?
Worldwide urbanization? Waste management or infrastructure
rehabilitation? Something military or medical? Once again, minerals
and energy? Or would it be shoreline protection as the ocean level rises?

The only certainty is that the industry will remain cyclical, declining
when most industries merely slow in growth. All other forecasts are
hazardous. After all, as late as 1970 the US National Export Council
wrongly saw fewer prospects for construction exports to the Middle
East than to Latin America. A committee chaired by Robert Fluor and
representing Bechtel, Parsons, Lummus, Foster Wheeler, and other
giants forecast a 1980 volume of only about $23 billion (1982 dollars)
for the Middle East and South Asia, which compares with the $65
billion that occurred (Report of the Industry Committee on Engineering and Construction Services, 1970). Later the boom had already
peaked when an otherwise sagacious observer saw no end in sight:
'One of the most widespread misconceptions regarding the Arab construction industry is that it consists of a fleeting boom' (Zahlan, 1983,
p. 249). By 1985 construction and related industries were the only ones
in absolute decline in the Middle East ('Growing Pains: the Gulf Co-
operation Council Countries: A Survey', *The Economist*, 8 February
1986). Any prediction attempted here would fare no better.

Wherever the boom occurs, it appears likely for reasons given
above, that firms from latecomer (NIC) countries will first enter the
most openly competitive parts of the market. The most energetic ones
may ride out storms together with those from advanced countries, but
no secular trend to displace the advanced – as happened with textiles,
shoes, and shipbuilding – seems in the cards. The contestable part of
the market will be lost to local firms, while the advanced countries will
keep what is most elaborate with their lead in finance and both data-
based and physical technology. Specific future techniques that will
dominate drafting rooms and sites are as unforeseeable as the

regions and product lines that will have called for them. Surely some good ideas will be met with undue pessimism, and some pipedreams will mislead many as inspired common sense. Fluctuations and discontinuity of work cannot be avoided, and inevitably they will hamper systematic technological progress.

Bibliography

ABEMI, Brazilian Association of Industrial Engineering, *Annual Reports* (Sao Paulo).

Aburish, Said K. (1985), *Pay-Off: Wheeling and Dealing in the Arab World* (London: André Deutsch).

Achilli, Michele and Khalid, Mohamed (1984), *The Role of Arab Development Funds in the World Economy* (New York: St Martin's Press).

Allen, Bruce T. (1976), 'Average Concentration in Manufacturing, 1947–1972', *Journal of Economic Issues*, vol. 10, no. 3, pp. 664–73.

Almeida, Julio Sergio Gomes de, and Ferraz Filho, Galeno Tinoco (1983), 'Estado, Politica Economica e a Constituicao de Grande Engenharia Brasileira', in Julio Sergio Gomes de Almeida, (ed.), *Estudos Sobre a Construcao Pesada no Brasil*, Universidade Federal de Rio de Janeiro-Instituto de Economia Industrial, Relatorio de Pesquisa 2 (Rio de Janeiro).

American Consulting Engineers Council (1986), 'Statement on Foreign Operations Funding before Subcommittee on Foreign Operations, House Committee on Operations', mimeographed (Washington, DC: ACEC).

Annibal Villela Consultoria Economica s/c Ltda (1981), Construcao Civil: Estudo Consolidado do Mercado Interno' (A); 'Mercado Externo para obras e servicos de Engenharia e consultoria', (B); 'Bases para a definicao de uma politica integrada para o setor de construcao civil', (Rio de Janeiro) (C), mimeo.

Associated General Contractors of America (1985, 1986, 1987), *International Construction Newsletter* (Washington, DC).

Banba, Tetsuharu (1982–4), 'The Overseas Construction Basic Problems Committee and its Interim Report', in Construction Industry Promotion (ed.), *Monthly Report of Construction Labour and Materials* (Tokyo).

Barrie, D. S. and Paulson, B. D. (1978), *Professional Construction Management* (New York: McGraw-Hill).

Behring, K., Gluch, E., and Russig, V. (1982), Entwicklungstendenzen im deutschen Auslandsbau, (Berlin–Müchen: Duncker & Humblot).

Bollinger, R. (1985), 'Auslandsbau 1982/85', *Bauwirtschaft*, vol. 51/52, pp. 1779–1922.

Brander, James, and Spencer, Barbara (1984), 'Tariff Protection and Imperfect Competition', in H. Kierzkowski (ed.), *Monopolistic Competition and International Trade* (Oxford: Oxford University Press).

Buckley, P. J. (1983), 'New Theories of International Business: Some Unresolved Issues' in Casson, M. (ed.), *The Growth of International Business* (London: Allen & Unwin).

Buckley, P. J. and Casson, M. (1976), *The Future of the Multinational Enterprise* (London: Macmillan).

Business International (1982), 'US Companies Fight to Save Ex–Im Bank from Budget Knife', 14 May (New York).

Business Week (1976), 'Where the Constructors Strike it Rich', 23 August (New York).

Capt, J., and Edmonds, G. A. (1977), *Study of Small Contractors in Kenya*, World Employment Programme, Working Paper, December (Geneva: International Labour Office).

Cassimatis, Peter J. (1969), *Economics of the Construction Industry*, Studies in Business Economics, No. 111 (New York, National Industrial Conference Board).

Casson, M. (1979), *Alternatives to the Multinational Enterprise* (London: Macmillan).

Caves, Richard E. (1982), *Multinational Enterprise and Economic Analysis* (Cambridge: Cambridge University Press).

Chaves, Marilena (1985), 'A Industria da Construcao no Brasil: Desenvolvimento, Estrutura e Dinamica', unpublished dissertation (Instituto de Economia Industrial da Universidade Federal do Rio de Janeiro).

Clough, Richard H. (1975), *Construction Contracting* (New York: Wiley–Interscience).

Cummings, Thomas (1984), 'Transorganizational Development', *Research in Organizational Behavior* (New York: JAI Press).

Dahlman, Carl J., and Cortes, Mariluz (1984), 'Mexico', in Sanjaya Lall, (ed.), *Exports of Technology* in *World Development* (May–June), pp. 601–24.

Daily Journal of the Construction Industry, Nikkan-Kensetsuklgyo-shinbunsha (Tokyo).

Demacopoulos, Alexander, and Moavenzadeh, Fred (1985), *International Construction Financing*, Technology and Development Program, Massachusetts Institute of Technology (Cambridge, Mass.).

Dixit, Avinash K. (1986), 'Trade Policy: An Agenda for Research', in Paul R. Krugman (ed.) *Strategic Trade Policy and the New International Economics* (Cambridge, Mass.: MIT Press), pp. 283–303.

Draper, William H., III (1985), 'Statement before the Committee on International Finance, US House of Representatives', October (Washington, DC).

Drewer, Stephen (1980), 'Construction and Development: A New Perspective', *Habitat International*, vol. 5, nos. 1–2.

Dunning, H. H. (1981), *International Production and the Multinational Enterprise* (London: Allen & Unwin).

Edmonds, G. A. (1979), 'The Construction Industry in Developing Countries', *International Labour Review*, vol. 118, no. 3.

Edmonds, G. A., and Miles, D. W. (1984), *Foundations for Change: Aspects of the Construction Industry in Developing Countries* (London: Intermediate Technology Publications).

Engineering News-Record (ENR), a number of issues, especially the annual surveys of US and international top contractors and design firms.

Ferreira, C. E. (1976), *Constrcao Civil e Criacao de Empregos*, Funacao Getulio Vargas (Rio de Janeiro).

FUNCEX (1982), 'Estudos Sobre a Exportacao de Servicos' mimeo,

Fundacao Centro de Estudos do Comercio Exterior, Project XCVI (Rio de Janeiro).

Fundacao Joao Pinheiro (1984), *Deagnostico Nacional da Industria da Construcao*, Diretoria de Projectos I, vols 1, 2, 5, 10, 12, 13, 14, 16, 18, and 19 (Belo Horizonte).

Gibson, Paul (1978), 'Another Domino is Falling', *Forbes*, vol. 122, no. 13, pp. 27–9.

Gluch, E. (1982), 'Beschaffung von Auslandsauftragen deutscher Bauunternehmen', *Ifo-Schnelldienst*, vol. 4, pp. 9–13.

Grandi, Sonia Lemos (1985), 'Desenvolvimento da Industria da Construcao no Brasil: Mobilidade e Acumulacao de Capital e da Forca de Trabalho', unpublished dissertation, Departamento de Ciencias Sociais, Faculdade e Filosofia, Letras e Ciencias Humanas da Sao Paulo, Universidade de Sao Paulo.

Grossman, Gene M. (1986), 'Strategic Export Promotion: A Critique', in James Krugman (ed.), *Strategic Trade Policy and the New International Economics*, pp. 47–68.

Halpin, Daniel W. (1986a), 'Tasks 1/2: Technology in Architecture, Engineering, and Construction', report for the Office of Technology Assessment, United States Congress (Washington, DC).

Halpin, Daniel W. (1986b), 'Task 3: Technology in Architecture, Engineering, and Construction', report for the Office of Technology Assessment, United States Congress (Washington, DC).

Hannan, Michel, and Freeman, John (1977), 'The Population Ecology of Organizations', *American Journal of Sociology*, vol. 82, no. 5 (March) pp. 929–40, 946–9, 955–64.

Hillebrandt, Patricia M. (1985), *Economic Theory and the Construction Industry*, 2nd edn (London: Macmillan).

Hippo, Yasuyuki (1983), *The Construction Industry in Japan: A Survey* (Bombay: Asian Productivity Organization).

Hufbauer, G. C. (1975), 'The Multinational Corporation and Direct Investment', in Peter Kenen (ed.), *International Trade and Finance: Frontiers for Research* (Cambridge: Cambridge University Press) pp. 253–319.

Hymer, S. H. (1976), *The International Operations of National Firms: A Study of Direct Foreign Investment*, based on 1960 MIT Doctoral Dissertation (Cambridge: MIT Press).

Iijima, Tadashi (1982), 'Short History of Overseas Construction Works of Japan' (serial), in Construction Industry Promotion Fund (ed.), *Monthly Report of Construction Labour and Materials*, June–September.

International Engineering and Construction Industries Council (1986), 'IECIC Paper on GATT', processed (Washington, DC), March.

International Engineering and Construction Industries Council (1983), *Rebuilding US Architectural, Engineering and Construction Services Exports: Testimony before the House Foreign Affairs Subcommittee on International Economic Policy and Trade, 24 May 1983* (Washington, DC: IECIC).

Jarboe, Kenan Patrick (1987), 'International Competition in Engineering and Construction', *International Competition in Services* (Washington, DC: Office of Technology Assessment).

Krugman, Paul R. (ed.) (1986), *Strategic Trade Policy and the New International Economics* (Cambridge, Mass.: MIT Press).

Lall, Sanjaya (1984), 'India', in Sanjaya Lall (ed.), *Exports of Technology by Newly-Industrializing Countries*, Special Issue of *World Development*, May–June, pp. 535–65.

Lange, Julian E., and Mills, Daniel Quinn (1979), *The Construction Industry: Balance Wheel of the Economy* (Lexington, Mass.: Lexington Books).

Lucas, Chester L. (1986), *International Construction Business Management: A Guide for Architects, Engineers, and Contractors* (New York: McGraw-Hill).

Mascaro, Lucia R. de, and Mascaro, Juan Luis (1980), *A Construcao na Economia Nacional*, Editora Pini Ltda (Sao Paulo).

McQuade, Walter (1980), 'An Expatriate Builder's Changing Frontiers', *Fortune*, vol. 102, no. 1, pp. 96–102.

Ministry of Construction (ed.) (1981–6), *A Year Book of Construction Statistics* (Tokyo).

Ministry of International Trade and Industry (ed.) (1986), *Introduction to Export Insurance* (Tokyo).

Moavenzadeh, Fred, and Hagopian, Frances (1983), *Construction and Building Materials Industries in Developing Countries* (Vienna: United Nations Industrial Development Organization).

Murphy, Kathleen J. (1983), *Macroproject Development in the Third World* (Boulder, Colo.: Westview Press).

National Constructors Association (1985, 1986), *NCA Newsletter* (Washington, DC: NCA).

Neo, Roland Bah (1975), *International Construction Contracting* (London: Gower Press).

Nihonkeizaishinbun-sha (ed.) (1985–6), 'Industrial Review of Japan 1975–1984', *Japan Economic Almanac* (Tokyo).

Ofori, G. (1981), 'The Construction Industry in Ghana', World Employment Program Working Paper (Geneva: International Labor Office).

Overseas Construction Association of Korea (OCAK) (1984), *Overseas Construction – Civilian White Paper* (Seoul).

Overseas Construction Association of Japan, Inc. (ed.) (1985), *Thirty Years of OCAJI* (Tokyo).

Pearce, J. (1980), 'Subsidised Export Credit', *Chatham House Papers*, no. 9 (London: Royal Institute of International Affairs).

Pinyan, Charles T. (1986), 'Foreigners step up U.S. Invasion', *ENR*, 27 November, pp. 12–13.

Price Waterhouse (1985), *The Contribution of Architectural, Engineering and Construction Exports to the U.S. Economy* (Washington, DC: International Engineering and Construction Industries Council).

Rhee, W. R., Ross-Larson, B., and Pursell, G. (1984), *Korea's Competitive Edge* (Baltimore: Johns Hopkins University Press).

Riedel, Jürgen (1983), *Global Prospects for the Development of the Construction and Building Materials Industry* (Vienna: United Nations Industrial Development Organization).

Riedel, Jürgen, and Schultz, Siefried (1978), *Bauwirtschaft und Baustoffindustrie in Entwicklungsländern* (Munich: Weltforum Verlag).

Riedel, J., and Schultz, S. (1980), 'Construction and Building Materials Industry in Developing Countries', *Economics*, vol. 21 (Tübingen).

Riedel, J. (1986), *Planning of the Construction Industry in Developing Countries* (Nairobi: United Nations Centre for Human Settlements (UNCHS)).

Rugman, A. M. (1981), *Inside the Multinationals: The Economics of Internal Markets* (London: Croom Helm).

Sercovich, Francisco Colman (1984) 'Brazil', in Sanjaya Lall (ed.), *Exports of Technology by Newly-Industrializing Countries*, pp. 575–99.

Seymour, Howard (1986), 'International Investment in the Construction Industry' Ph.D. thesis, University of Reading, and forthcoming (London: Croom Helm).

Shim Ui-Sup (ed.) (1984), *Korean Construction in the Middle East* (Seoul: Bub Mun Sa).

Solo, Robert A. (1975), *Organizing Science for Technology Transfer in Economic Development* (East Lansing: Michigan State University Press).

Stalson, Helena (1985), *U.S. Service Exports and Foreign Barriers: An Agenda for Negotiations* (Washington, DC: National Planning Association).

Stephenson, John C. (1984), 'Technology Transfer by the Bechtel Organization', in Ronald K. Shelp, John C. Stephenson, Nancy Sherwood Truitt and Bernard Wasow, *Service Industries and Economic Development: Case Studies in Technology Transfer* (New York: Praeger).

Strassmann, W. Paul (1970), 'The Construction Sector in Economic Development', *Scottish Journal of Political Economy*, vol. 17, no. 3 (November) pp. 391–409.

Telles, Pedro C. da Silva (1984), *Historia da Engenharia no Brasil* (centuries XVI to XIX) (Rio de Janeiro: Livros Tecnicos e Cientificos Editora).

Thode, D. (1978), 'Ingenieurbau im Ausland', *Baupraxis*, vol. 7, pp. 13–16.

Tucker, Richard (1985), 'CII Project Overview', presentation to the First CII Annual Meeting (Austin: University of Texas).

Turin, D. A. (1978), 'Construction and Development', *Habitat International*, vol. 3, no. 1–2, pp. 33–45.

US Army Corps of Engineers (undated) *The Middle East Division* (Winchester, Virginia: US Corps of Engineers).

US Bureau of the Census (1975), *1972 Census of Construction Industries* (Washington, DC: US Government Printing Office).

US Department of Commerce (1984), *A Competitive Assessment of the US International Construction Industry* (Washington, DC: US Government Printing Office).

US International Trade Commission (1982), *The Relationship of Exports in Selected Service Industries to U.S. Merchandise Exports* (Washington, DC: USITC).

US Trade Representative (1983), 'U.S. National Study on Trade in Services', mimeographed (Washington: Office of the US Trade Representative), December.

Wells, Jill (1985), 'The Role of Construction in Economic Growth and Development', *Habitat International*, vol. 9, no. 1, pp. 55–70.

Wells, Jill (1986), *The Construction Industry in Developing Countries: Alternative Strategies for Development* (London: Croom Helm).

Westphal, Larry, Rhee, Yung W., Kim, Linsu, and Amsden, Alice H. (1984), 'Republic of Korea', in Sanjaya Lall (ed.), *Exports of Technology by Newly-Industrializing Countries*, pp. 505–33.

Williamson, Oliver E. (1981), 'The Modern Corporation: Origins, Evolution, Attributes', *Journal of Economic Literature*, vol. 19 (December) pp. 1537–68.

Williamson, Oliver E. (1985), *The Economic Institutions of Capitalism: Firms, Markets, and Relational Contracting* (New York: Free Press).

World Bank (1984), *The Construction Industry: Issues and Strategies in Developing Countries* (Washington, DC: The World Bank).

Young, Crawford (1982), *Ideology and development in Africa* (New Haven, Conn.: Yale University Press).

Zahlan, A. B. (1983), *The Arab Construction Industry* (London: Croom Helm).

Index

Hifab 177
Hillebrandt, P. M. 43, 232
Hippo, Yasuyuki 232, 240 (Joint
  contribution) 59–85
Hitachi 67, 153
Hochtief AG 123, 127, 217
Hoffman & Sönner 162, 168
Hogjard & Schultz 162, 167–8
Holzman 178
Hong Kong 27, 61, 78, 84, 158, 211
hospitals 38, 115, 129, 143, 163
hotels 61, 74, 163, 170, 174, 175
housing 38, 118, 161, 172, 209
Hume-Balken 176
hydraulic engineering 203
hydroelectric projects 2, 60, 90, 180, 187,
  190, 213
Hymer, S. H. 228
Hyundai Group 143, 144, 145–7, 149,
  230

Iceland 169
IECIC 47
IFAWPCA 61, 143
IFO 131–2
immersed tunnels 168
Impianti 223
Imprefeal 99
Impregilo 90, 93, 230, 244
Impresit 90, 96, 230
INCO 208
India 112, 121, 174, 211, 225
indigenization 79, 81, 215
Indonesia 64, 78, 111, 143, 158
industrial plants 171, 229
industrialized building methods 161–2,
  172
insurance 83, 84–5, 101, 102, 156, 170,
  203, 207, 214, 235, 239–40, 242
Interamerican Development Bank 187,
  245
international
  Construction Week 6
  development associations 245, 246
  development banks 245–7
  Eng. and Construction Industries
    Council 47
  finance 79, 158, 225
internationalization 173, 181, 186
Iran 83, 95
Iran Iraq War 62, 63, 83, 94, 106, 189,
  244
Iraq 114, 147, 168, 185, 189, 229
IRB 196

IRI 97, 238
irrigation schemes 129, 163
Israel 123, 211
Italy
  African projects 213
  aid financing 91, 102, 227
  attitude to bribery 238
  Bosporus Bridge contract 95, 216
  bureaucracy 101–3
  business strategy 92–7
  collaboration with foreign firms 94,
    169, 203
  colonial ventures 89
  consortia 229–30, 237
  counter trade 95–6
  emigration 102–3
  equipment 99
  European contracts 6
  expatriate workers 91, 98–9
  export insurance 101–2
  Fascist regime 94
  financing 190, 216, 244
  government policy 100–3, 216
  International Construction Statistics
    87 Table 4.1
  labour laws 98–9
  major groups 93–4
  Ossola law 101
  specialization 229–30
  unification (1870) 102
Itamaraty 188
Ivory coast 114
Iwo Jima 143

Japan
  artificial island 52
  attaches 84, 236
  'big five' contractors 78
  business strategy 59–60, 66–81, 92–6,
    227
  competitiveness 74, 158
  consortium with Italians 237
  Construction Industry Law 61
  construction market 1
  contractors 60, 61–4, 67 Table 3.2, 69
    Table 3.3, 76, 78
  counter trade 85, 238
  designers 73, 232
  development aid 70, 73, 83, 235
  disincentives 71 Table 3.4, 72–6
  domestic demand 68
  'dumping' accusations 150
  Edo period 66
  European subsidiaries 70